Cryptic Science™

The Secret of Groom Lake

By
S. Arthur Hart

An
ADVENTURE BOYS™
Action Story

The Secret of Groom Lake

An
ADVENTURE BOYS
Action Story

Published by Adventure Boys Company ™
A Division of Adventure Boys, Inc.™
Seattle, Washington

Creative Director: Mark Jacobsen
Art Director/Designer: Jim Dixon
Cover Illustration: Shane White
Illustrated by: Shane White
Additional Illustrations and Content by: Jim Dixon,
Blake Manning and Drew Woods
Creative Advisor: Steve Kelley

ISBN 10: 0-9796392-0-4
ISBN 13: 978-0-9796392-0-3

First Edition May 2007
10 9 8 7 6 5 4 3 2 1

Printed in the United States of America

www.adventureboys.com

CRYPTIC SCIENCE™

Cryptic Science™ is a historical science fiction series that follows the life of 15-year-old Enrique "Ric" Lopez. Set in Groom Lake, Nevada in the 1950s, each action-packed Cryptic Science™ story tests Ric's abilities, challenges his loyalties and takes him one step closer to becoming a man.

Cryptic Science™ is one of six historical fiction serials that include Wild Boys Adventures™ (1860s westerns), Detective Mysteries™ (1930s mysteries), Blue Squadron™ (1940s aviation), Spy Racer™ (1960s racing/espionage), and Treasure Raiders™ (1970s action adventure).

THE WORLD OF ENRIQUE "RIC" LOPEZ

You might not believe in UFOs and frightening space aliens, but as the agents from the United States Special Operations of Covert Security will tell you, "There are a lot of things the public doesn't know about." Fifteen-year-old Enrique "Ric" Lopez is about to discover just how weird things can get when the veil of mundane life is pulled away.

The year is 1956, and the world is embroiled in the Cold War, an escalation of tensions between the United States and the Soviet Union. Each nation vies to develop weapons and technology that will give them the advantage over the other. In the United States, President Dwight D. Eisenhower authorized the creation of Area 51, a top-secret research facility located on the Groom Lake Air Force Base, deep in the Nevada desert. The mission of Area 51 is to explore the boundaries of science and create new and exotic technologies for the betterment of America and her allies.

When Enrique Lopez's father is reassigned to Groom Lake, it could not have happened at a worse time. The passing of Enrique's mother has left his father distant and Enrique feeling alone and bitter. As the only Latino kid on base, Enrique is subject to isolation, prejudice and even ridicule from many of his peers.

But unbeknownst to Enrique, Area 51 holds a mysterious secret; a secret that if exposed would shake the very foundations of society. A secret, which in the wrong hands, could mean the destruction of the United States. A secret Enrique is destined to discover.

What amazing adventures await Enrique Lopez beneath the sands of Area 51? Read on, if you dare, and discover ... The Secret of Groom Lake!

THE CHARACTERS

Thomas "Tommy" James Reilly

The son of an army sergeant, Tommy Reilly moved from army base to army base for as long as he could remember. Quiet and reserved, Tommy's fondness for cowboy shirts and hats had set him apart from most other army kids. However, when Tommy arrived at Groom Lake Army Base he immediately struck up a friendship with Carol Johansson, who took an interest in Tommy's knowledge of cowboys and the Old West. Tommy has always enjoyed the outdoors, especially his time working on his uncle's ranch.

Carol Jean Johansson

Unlike most of her peers at Groom Lake base, thirteen-year-old Carol Johansson has taken a voracious interest in her studies, especially science. As the daughter of Groom Lake base commander, General George S. Johansson, Carol enjoys a certain amount of privilege and liberty in regards to the secret operations on the base, including access to some of the laboratories in the secret Area 51 sector. Carol's growth spurt over the last year has brought some unexpected changes both in Carol's appearance and self awareness.

Captain Roy Davis

Captain Davis was Groom Lake's top military test pilot – until an unexpected medical condition prevented him from flying. Unable to perform his normal duties, Captain Davis was reassigned to an animal training program, preparing various animals for possible combat and field operations.

Esteban Lopez

Enrique's father is a highly skilled airplane mechanic working at the Groom Lake test facility. Esteban loves his son dearly, but the recent death of his wife has weighed heavily on his heart, and has made him somewhat distant.

Professor Gloria Appleby

One of the world's top scientists, Professor Appleby is involved in some of the nation's most secret projects. A free spirit and forward-thinking libertarian, Professor Appleby constantly seeks to use science and technology for the betterment of mankind. Her passion for life extends into the arts, especially music from various cultures.

Agent Jones

Nobody knows much about Agent Jones. Nobody, that is, except for high-ranking government and military officials. As a secret operative, Agent Jones considers himself to be somewhat "above the law." His real name is unknown. His mission is unknown. All that's known, really, is that he works for the U.S. government …

TABLE OF CONTENTS

For Liam, Anna, Brenna,
Dillon and Mari, and
the adventures ahead.

PROLOGUE

Her first conscious thought was the realization that she could not move.

At first she thought that she must not be awake. But she knew differently. A bitter, medicinal taste filled her mouth. She tried wiggling just a finger but could not. Her body tingled as if every muscle were asleep.

All was dark and she felt the pull of drying tears on her eyelids. Had she been crying in her sleep?

"Ah, I see you are awake. Excellent." The male voice came from somewhere behind her. It echoed throughout the chamber. It was soft yet resonant; almost melodic.

"Where am I? Can you help me?" Her voice was hoarse. Her pleas came out as a whisper.

Lights rose to a dim glow all around her. She was lying upon a stainless steel table; surrounding the table was a cylindrical pane of glass. It was like she was lying in a giant test tube except that there were several small holes bored into the glass to let air in. She was fully clothed in her flowered red dress and laboratory coat. There were no restraints that she could see, but she was trapped inside the glass enclosure just the same.

Around her were the tools and devices used to conduct medical experiments on animals. This was a medical laboratory. As a scientist, she hated these places, loathed that they were a part of the scientific process.

She tried to look beyond her immediate surroundings but all was dark. Where was she? The room was not at all like the well lit labs on the air force base where she worked. Beyond the shadows it looked like the walls were dirt and rock but she could not be certain.

Was she still on the base? If so, one thing was certain: she had never been in this laboratory before.

A figure stepped from the shadows and into her line of sight. It was a man, tall with bony, square shoulders, yet thin of build. He was dressed in a lab coat and he wore upon his head a surgical mask and cap.

The man leaned forward, pressing his masked face against the glass; his mouth was placed directly over one of the air holes. Professor Appleby could smell garlic on the man's breath, hear a shallow rasp in his exhale.

"I would like to know about Project Nephilim."

Nephilim! She felt the cold hand of the military's conditioning clamp down on her throat. She labored for breath.

"Please … " The words were hard to speak. "I can't tell you anything … "

"Yes. I know that. And I am sorry for it."

He did not sound sorry at all. The man pulled on a large lever that was attached to the base of the bed. There was a loud, scratchy electrical hum, and after a moment the bed began to tilt upright from a horizontal to a vertical position. When the bed was at a 75-degree angle the man in the mask pulled the lever again and the bed stopped rotating. The man then reached out and unfastened a latch at the top of the glass enclosure above her head.

"What are you doing?" she asked.

The lid of the glass enclosure was attached with hinges. The man in the mask flipped the lid open. She felt a rush of cold damp air.

The man in the mask then brought a small silver device to his mouth and let out a sharp whistle.

She heard the scraping of rocks, and several pebbles fell into the glass tube enclosure. Then what sounded like two silk sheets being rubbed together. Looking up into the gloom, her mind reeled.

Descending upon a thick rope of webbing was a creature unlike any she had ever seen. It was an abomination of nature. Its body was that of a spider but its head was humanlike. Its round yellow eyes bulged and were filled with malice. Its teeth were sharpened to points and clacked noisily against each other. The creature was slightly larger than a house cat. Its eight legs were long and spindly, and covered in coarse, barbed hair. The creature used these legs to control its descent. As the creature neared the opening of the glass enclosure, she froze. Her eyes locked with those of the creature. Intelligence gleamed in the creature's foul, baleful eyes. And she screamed!

The creature cracked a wicked smile that split its wide, round face. It then reared back its head and made a hideous howl, and as it did, two canine fangs slowly distended from its gums, forming two elongated spikes.

"You'll find that the poison of this creation, although painful, is quite useful in eliciting the truth from those unwilling to give it."

The creature turned to its master and asked in a vulgar, raspy voice. "For me?"

"Yes," replied the man in the mask. "But make it last. I have questions for our dear Professor."

CHAPTER 1

An Unlikely Place

Nye County, Nevada
September 26th, 1956

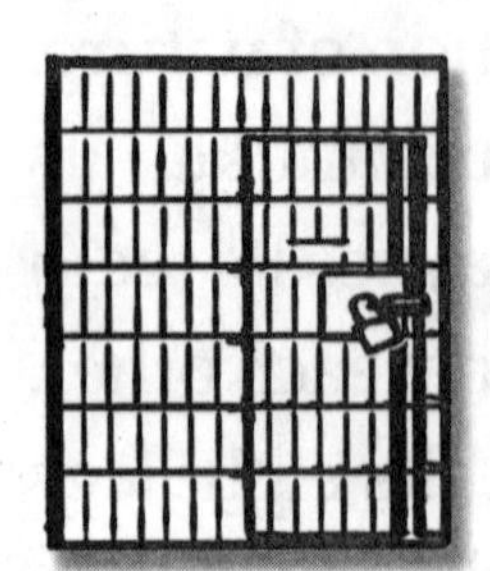

The town of Pahrump, Nevada was like many small towns in the western United States in 1956. There was a post office, a grocery, a feed store, three churches, four saloons, a hotel and, of course, a town jail. The Pahrump jail had not changed a whole lot since it was built in 1891. Featuring just one jail cell of brick and wrought-iron bars, the small structure was a true western pokey, which rarely saw more than one occupant in any given week. Today was a different matter, however.

Fifteen-year-old Enrique "Ric" Lopez sat in the corner of the Pahrump jail with his back against the bars. His Levi jeans were dusty, his white t-shirt dirty and his mood sullen.

A whole lot of trouble over a fifteen-cent movie ticket.

Most fifteen-year-old boys who found themselves sitting in a dusty Nevada jail would be powerfully worried about the reaction of their father once they got home; those reactions mostly involved a leather belt and the boy's behind. But the truth of the matter was, Enrique didn't have much concern about the belt. This was because Enrique believed his dad didn't care one way or another whether he was in a jail or in a church.

Across from Enrique, sitting on the only bench, were two hard-looking men. Although neither appeared much older

than thirty, the pain of war could be read in each man's eyes and the premature lines on their faces.

Veterans of World War II, thought Enrique.

Both wore tired, dusty black leather jackets and sported the "Jelly Roll" hairstyle – meaning their hair was combed up and forward on both sides and then brought together in the middle of the forehead, ending with a little curl. The style was held in place by a healthy dose of Brylcreem.

The man to Enrique's right was big – easily six-feet-four and three hundred pounds. His head was anvil shaped, his lips were thick and the ridge of his brow, knuckled.

This one's not the brains of the two, that's for sure, thought Enrique.

That distinction would fall on the man to Enrique's left. He was all business; a James Dean look-alike, complete with toothpick and attitude.

Anvil Head leaned over to the man with the toothpick and whispered something in his ear. He then gave Enrique a roguish stare.

"Yo, Jose. Whatcha in for? Steal a tricycle or somethin'?" The big brute elbowed his partner and grunted a laugh.

Enrique took a moment, pulled a comb from his pocket and raked it through his jet-black hair. Although Enrique's pompadour hairstyle was not nearly as "boss" as his cell mates', it was no flattop or crew cut, which was what most "good" kids his age wore.

"My name is not Jose. It is Enrique; and I was arrested for running an illegal business."

Both Anvil Head and Toothpick found this amusing.

"Is that right?" asked Anvil Head.

"Si. I was charging people a nickel to watch your mother dance on an apple box in the park."

It took a moment before the insult sank in, but when it did, Anvil Head looked as serious as a freight train, and just as dangerous.

"What did you say?"

"Oh, yes," continued Enrique casually, "the sheriff arrested me. He said it was illegal to parade a chimpanzee in public without a leash."

Once again it took a moment for the insult to hit home, but when it registered, Anvil Head exploded. In one fluid motion he shot out of his seat, grabbed Enrique by the neck and hoisted him high off his feet by the throat. Behind Anvil Head, Toothpick was laughing so hard he was slapping his knee.

"Are you callin' my mother a monkey?"

"No, no," said Enrique in a strangled voice. "Chimpanzees are apes, not monkeys."

Anvil Head cocked one arm back and was about to deliver his ham-sized fist into Enrique's fifteen-year-old-nose.

"Why you little … "

Fortunately for Enrique, Toothpick had not taken his insults so personally.

"Take it easy, Max," said the man with the toothpick, "I like this kid."

Anvil Head took a long look at Enrique, who was now turning blue from the choke hold.

"Let him be, Max," said Toothpick. "I like this kid, he's funny."

The humor of the boy's insults apparently began to register on the big man, who slowly cracked a gap-toothed grin. He let Enrique drop to his feet and shook him a bit as if to get the air back in him.

"Yeah, kid, I guess that's a good one. Youse pretty funny."

Somewhat breathless from being choked, Enrique found his former spot in the corner next to the bars.

The man with the toothpick took a seat on the bench, leaned forward and offered Enrique one of his Lucky Strikes.

"No thanks," said Enrique. "I promised my mom I wouldn't smoke."

"Your old lady don't need to know," said Toothpick.

"Oh , she'll know alright … " said Enrique as he held out

the crucifix that was around his neck. "She passed away a little over a year ago."

Toothpick put his smokes away.

"Sorry to hear that, kid. My name's Sonny, the big fella is Max."

"Nice to meet you," said Enrique. "Are you guys in a gang or something?"

"I'd like to think of it more as a club," said Sonny. "A biker club. We ride motorcycles."

"No kidding? My dad gave me a '46 Harley Flathead to repair." Enrique paused for a moment.

"He promised we'd rebuild the bike together. What a mistake it was to believe that load. It's still sitting in parts at the air force base. He may as well have given me a pile of junk for all it means to him." Enrique picked a pebble off the ground and dejectedly threw it across the cell.

"A '46?" said the big man, beaming his toothless grin. "Right on man, dat's what I ride."

Sonny scooted off the bench and sat down next to Enrique.

"Is your dad the reason you ran away from the base?"

"How did you know?" asked Enrique.

"I was an army brat myself. I've been there, man."

"Oh yeah, but I bet you weren't the only Chicano kid on the base."

"No, but I had plenty of other problems," said Sonny. "But that's what life's about, man. What fun would this be if life didn't throw a few potholes in your road every now and then? Those potholes give you drivin' skills, man, they make you swerve, find your groove. You know what I'm sayin'?"

"Yeah, I guess so," said Enrique.

The jailhouse door banged open and Pahrump's cowboy

sheriff lumbered in, cell keys jangling from his belt. Buster Hank, a large man swinging a decent-sized gut in from of him, approached the cell. Enrique was trying not to like the guy. But Buster had an infectious grin and a western drawl. He also wasn't a big fan of the military base, and that scored a few points with Enrique as well.

"Enrique Martinez Lopez. You've made bail. Come on out. You other boys will be getting out soon enough."

Enrique was slow to get up, so Max crossed to him, hefted the boy off the floor and gave him several friendly thumps on the shoulder.

"You got guts kid. I've not met many pugs that would say what youse said to me. That counts for somethin' in my book."

Enrique managed a smile. "Thanks, you big ape."

Sheriff Hank led Enrique out of the cell and toward the door.

"Yo, kid," Sonny called from behind the closed cell door. "My old man only gave me one thing worth a damn in his entire life, and it was a piece of advice. He said that life is going to be hard, but that doesn't mean you have to make it hard on yourself."

Sonny then tossed Enrique his leather jacket.

"For when you get that Flathead up and running."

"But, but —" Enrique stammered.

"I know it's beat up a bit, but it still has a good few miles left in it," said Sonny. "Time for me to get a new one anyway."

Enrique was speechless. It had been some time since anyone had given him anything, whether it was advice, a gift or fatherly company … and now he had found them all in the most unlikely place of all — a dusty jail in a sleepy little podunk called Pahrump.

CHAPTER 2

The Spooks

By the time Enrique could think of something to say, he found he had been escorted by Sheriff Hank to the sidewalk. Although the sun had been up for less than two hours, the promise of another scorching Nevada day was already in the works.

Waiting for Enrique in front of the jail was Captain Roy Davis. A legend on the base, he had been Groom Lake's top test pilot until an unexpected medical condition had recently grounded him. But Davis had taken it in stride, embracing his new role in a project involving animal training for possible combat and field operations. Enrique was a dog lover, which was how they met. Captain Davis was one of only a handful of people Enrique considered a friend.

Opening the door to his standard-issue jeep, Captain Davis gestured for Enrique to step in.

"Unbelievable," said Enrique with unsubtle bitterness. "He couldn't even come to pick me up? Unbelievable."

Enrique shoved his arms into his new leather jacket, thrust his hands into the coat pockets and stormed past Captain Davis.

"Look, Ric —" began Captain Davis.

"It's Enrique, not Ric, and I don't want to hear it!"

"Enrique, your father didn't want it to be this way."

"Yeah, right."

The captain grabbed Enrique firmly by the elbow. "Enrique,

I know you think your father doesn't care about you. He does. And I know it's hard for you to see it but there are other factors, other pressures your father is under."

"I get it! He's the Chicano engineer; the first Chicano with a top secret security clearance. Well, let me tell you something, Roy, there are more important things than work. Like family! You know who told me that? My mother."

Enrique turned away but Captain Davis pulled the boy abruptly back to him.

"Don't think for a minute that your father didn't love and cherish your mother. He did. You think that he's doing what he's doing for himself? Well, he's not."

Just then a black Chevrolet Bel-Air sedan pulled up in front of them. Even before the sedan had come to a complete stop, two men exited the car and moved toward Captain Davis and Enrique. These men, like the one remaining behind the wheel of the sedan, were dressed in black suits with white shirts and black ties. Their hair was close-cropped, their features angular. All wore black sunglasses and moved with the economy of trained combat men. Enrique had seen men like these back at the air force base. Soldiers spoke about them in hushed whispers. They came and went as they pleased, like ghosts in the night. Even General Johansson himself, commander of Groom Lake Air Force Base, granted these men a degree of liberty. Their identification cards said they were Special Operatives of Covert Security. The regular military simply called them "The Spooks."

Captain Davis stepped between Enrique and the two men. Enrique noticed that Captain Davis had assumed a *Kamae* stance. Kamae was an Aikido term for an alert posture of martial readiness. The captain was a first-dan black belt in

Aikido and Groom Lake's self-defense instructor.

"Enrique Lopez?" asked the man on the left.

Davis looked hard at one man, then the other. His eyes were tight, his breath even.

"Can I help you gentlemen?" he asked, his voice soft yet stern.

"Are you Enrique Lopez?" asked the spook on the right, his eyes fixed on Enrique.

"I am Captain Roy Davis, United States Air Force … "

The man on the left moved in and grabbed Enrique by the shoulder and wrist. The captain was catlike in his response. He dropped low and stepped in close, grabbing the man's wrist and wrenching it. At the same time he delivered a solid blow to the man's elbow. He quickly circled right, which wrenched the spook's elbow joint painfully, forcing him to one knee. Captain Davis quickly brought his knee up into the man's jaw, knocking him to the sidewalk, unconscious.

The man on the right made a move to grab Enrique in a neck lock but Enrique swiftly dropped and rolled out of the way, a move Captain Davis had taught him back at the base.

The captain then stepped forward and delivered a blow to the man's solar plexus, knocking the wind out of him. Circling behind the stunned agent, he wrapped one arm around the agent's neck, cutting off his wind.

"As I was saying, I am Captain Roy Davis, United States Air Force, Groom Lake. I am under direct orders from Groom Lake Base Commander General George S. Johansson to escort this civilian back to the base."

Captain Davis loosened his grip on the man's neck, allowing him to take a breath and speak.

"I am Agent Rice of the Special Operations Covert –"

"I know who you work for," said Captain Davis. The agent's face began to turn blue from the choke hold.

"Ugh ... a scientist has gone missing from the base ... " said the spook.

Captain Davis tossed the man aside. The man who had been knocked unconscious slowly rose to one knee. As he stumbled to his feet, the agent lurched to one side and accidentally bumped Enrique.

"We are returning to the base now," said Captain Davis. "If you wish to question me or Mister Lopez you may do so back at the base. However, Mister Lopez will be returning with me, as per my orders."

Neither man argued; instead they both straightened their suits and stepped back, allowing Enrique and Captain Davis to pass. Enrique paused just for a second and gave the unsteady one a tight-lipped smile.

"That looked like it hurt," said Enrique.

He felt the man's stare burning into the back of his neck all the way back to the jeep.

Soon after the jeep pulled away, another black Bel-Air four-door sedan, this one with darkened windows, crept to a halt next to the special operatives. The rear window rolled down. Sitting in the back seat was a thin, emaciated-looking man. His smallish, pear-shaped head was bald except for wispy, combed-back black hair around the temples. He wore a pencil-thin mustache and a silk ascot scarf around his neck. He smoked a cigarette with a filter, which he held between his thumb and middle finger.

One of the spooks leaned close to the window, and in Russian, spoke, "Спящий был размещен."

Unexpectedly and without any indication, the man in the car casually extinguished his cigarette on the spook's cheek, searing his flesh and causing him to stifle a yelp of pain.

"In English, you fool," said the man in the car. His voice was incongruously deep and rich considering his sallow, milky features.

"My apologies, Commander. The mission was a success. The sleeper has been placed."

"Excellent," said the commander. "Our mole has confirmed that Project Nephilim has been relocated to Area 51 at Groom Lake Base. Once we receive word on its precise location, we shall wake the sleeper."

"What is the sleeper's mission?" asked one of the spooks.

"That is not your concern," said the commander. "But rest assured it does not bode well for the Americans."

CHAPTER 3

Groom Lake

Groom Lake Air Force Base was located in southern Nevada and was one of America's most secure and secret facilities. It was nestled amidst a remote tract of land that was surrounded by an imposing line of hills, mountains and ridges. There was only one road leading from Groom Lake to the nearest highway, State Route 375. The first security checkpoint was miles from the official base perimeter, and there were three other security checkpoints along the base road before anyone reached the entrance gate. All four checkpoints were heavily armed and no one was permitted through without the proper identification. The reason for such rigorous security was that within the Groom Lake Base resided Area 51; a complex of super-secret scientific facilities conducting classified, experimental military research.

Enrique and his father, Esteban, had moved to Groom Lake shortly after his mother's death. His dad was an engineer who specialized in propulsion systems, and was recruited to work at Area 51. And although his father was not in the military, he was a military contractor and thus Enrique had spent nearly his entire life on one military base or another.

From the first moment Enrique set foot on Groom Lake, he knew it was different from any other base he had lived on. Its top secret nature meant that everybody who worked at Groom Lake lived at Groom Lake, whether they were civilian or military. Everyone was required to carry identification at all times. Area 51 was off limits to most base personnel, including Enrique, but he had heard talk of underground research facilities, camouflaged installations and super-secret bunkers. Whether his dad worked at any of these, Enrique did not know. His father was very tight lipped about his work.

Of course, Groom Lake Base also had the amenities that most bases had, including housing barracks, a commissary, gymnasium and a school.

Enrique couldn't stand school; at least the school at Groom Lake. It wasn't so much the school work – that was easy enough – it was the people. He could not count one friend among the sixty or so students who attended classes. Perhaps a friend would have talked him out of sneaking off the base the previous evening. A friend would probably have advised against sneaking into the movie theater, and a friend would surely have cautioned him against mouthing off to the theater owner, which had landed him in jail.

Well, that was then and this was now. And now he told himself he couldn't care less what trouble he had coming.

Sneaking off the base had been worth it. The base, his life, had become a prison and he'd needed out.

When the jeep pulled up to the first security checkpoint, Enrique noticed there was a full contingent of soldiers on gate duty; apparently the base had increased its security threat level. When they reached the fourth security checkpoint, both Captain Davis and Enrique were asked to step out of the jeep for a pat down search. When he emptied the pockets of his new leather jacket, Enrique discovered a large metal butane cigarette lighter.

That's weird, thought Enrique. Had that been there the whole time?

Enrique swore he had checked the pockets when he first put on the jacket. He wondered if the biker had intended for him to have the lighter, too. It was plated with an embossed metal eagle design on one side and a metal American flag on the other. The security guard flicked it open and checked to see if it worked. It did. The flame was bright and tall.

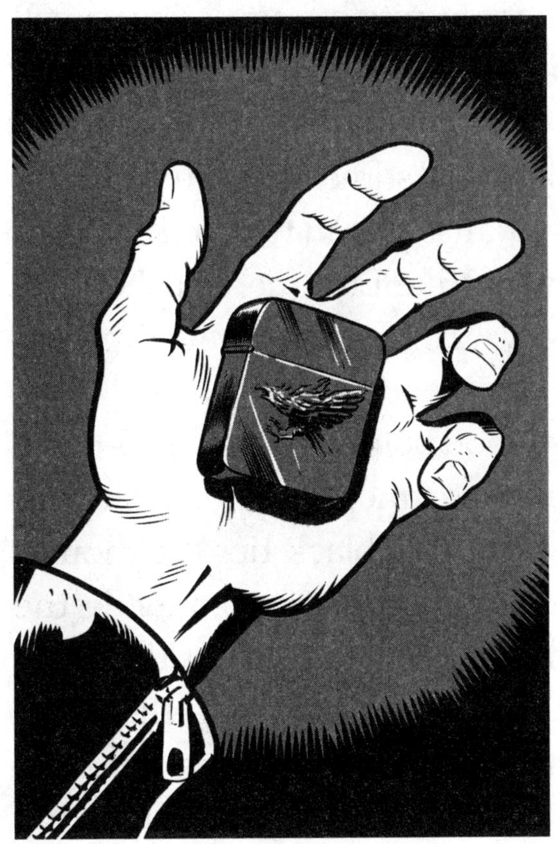

"Nice lighter," said the guard.

Captain Davis gave Enrique a disappointed look, as if to say, "So now you're smoking?"

Enrique was going to protest but then he thought, why bother? The guard returned the lighter to Enrique and both he and

Captain Davis proceeded into the base, stopping in front of the command building. Built out of large concrete blocks, it was painted the usual drab military gray, and had several communications antennas sprouting from its roof.

General Johansson, the commander of Groom Lake Air Force Base, was seated at his desk in his office. Enrique could see that he was busy with one of the Groom Lake scientists – a worker, in the same long white lab coat that his father often wore. Enrique went to take a seat in the reception area to wait until the general was free. Instead Miss Eckstein, the receptionist, conducted him and Captain Davis straight to the general's office. The general dismissed the scientist and waved them in. Mrs. Eckstein shut the door behind them.

General Johansson was a heavyset man. Although he appeared fat, Enrique knew from watching the general work on the heavy bag in the gym that, despite his age, he was well muscled beneath that uniform. The general was something of an enigma to Enrique. Several times the boy had found himself in front of the general due to some bit of mischief or trouble, and each time General Johansson had treated him with a degree of courtesy, which was unlike Enrique's teacher and school administrator, Mister Pan, who liked to berate him and shout him down.

Captain Davis and Enrique stood before the general's desk. From the shadows of the office two men stepped forward. Both wore black suits, white shirts and black ties. Although these were not the same men Enrique had met outside the Pahrump jail, he had no doubt that they worked for the same agency. The man on Enrique's left was older than the other spooks he had seen. And unlike all other spooks, this one wore a United States flag pin in his lapel, which Enrique took as a sign of authority.

This spook was not someone to mess with, Enrique thought.

General Johansson took off his black-rimmed glasses and set them on the desk.

"Ric, why did you leave the base yesterday?"

The general's voice was soft and not the least bit angry or threatening.

"I prefer Enrique, sir."

"Alright. Enrique, was there a reason why you left the base without authorization?"

"I don't know sir. I guess I needed to leave."

"That is not an answer, Mister Lopez." The remark came from the older spook, the one with the flag pin. His voice was mechanical but with an edge, like the noise of a knife being sharpened. He spoke his words slowly and precisely. Enrique's stomach tightened and his hands began to sweat. The spooks had that effect on people. The older spook stepped toward Enrique and removed his sunglasses, revealing black, intense eyes and a scar over his left cheek.

"What time did you leave the base, Mister Lopez?"

"Um … I'm not sure. It was just getting dark … "

"How did you exit the base without being observed?" asked the spook, who took another step closer to Enrique. The proximity and intense bearing of the spook made Enrique feel as if the walls of an inescapable trap were closing in on him.

"Um, it was … it was a truck … a produce truck … " Enrique stammered.

"So why did you leave the base hidden in a produce truck, Mister Lopez?" asked the spook, his voice louder.

"I um, I don't know, I … " He looked to the general but

found no help.

"Was it because your father had asked you to?" shot the agent.

"No."

"Was it because your father had given you something to take off the base?"

"No … I hardly see my father, we hardly speak … "

"Do not lie to us, Mister Lopez!" snapped the spook.

"I'm not lying! I … I wanted to see a movie but I got caught sneaking into the theater. That's all."

"You're lying, Mister Lopez," said the agent, "You're not telling us the real reason you fled the base."

"That's enough," said Captain Davis. "Excuse me, General, but you've read Enrique's personal file, you know the difficulties he's had since his mother's death, and the trouble he's been having with the other students. He needed a break. He wanted to see a movie. That's it. Boys his age do this all the time."

"Do boys his age also commit murder?" asked the Spook.

"At ease, Agent Jones," said the general in a firm voice. "No one knows if a murder has even taken place."

"Have you ever been inside Area 51, Mister Lopez?" asked the spook. "Specifically Alpha sector, sub-laboratory 14?"

"I said, at ease, Agent Jones!" barked the general.

"What's this all about, General?" asked Captain Davis.

"Professor Appleby is missing. There appears to have been a struggle in her lab. She is nowhere to be found. Security suspects foul play."

Enrique felt as if he had just been kicked in the stomach. The voices in the room became distant. The office began to swirl.

Professor Appleby missing? thought Ric. *Why, I just talked with her yesterday.*

The general took a chair and gently sat Enrique down.

"Enrique, I understand you and Professor Appleby were close, and that you saw her last night. Did she say anything to you? Did she discuss any problems she might be having?"

Enrique's mind was swimming ... was there something he knew that might help?

"No, no ... nothing," he said in a soft voice.

"It appears that some of the laboratory animals may have gotten loose," said the general. "Did you have anything to do with that?"

"No!" said Enrique sharply.

"Mister Lopez," began Agent Jones, "Did you not three weeks ago set free a number of laboratory animals?"

"Yes. But it was a prank. They were harmless and they were all returned to their pens." He glanced at the general. "You don't think I had something to do with the professor being missing, do you? Professor Appleby is my friend!"

His eyes burned and his muscles trembled. Captain Davis stepped in and took Enrique by the shoulders.

"General, Enrique has had a difficult night," said Captain Davis, "He needs some rest. And I think his father should be here if there are any more questions."

"Good enough," said the general. "Get some rest, Enrique, and we'll work through this tomorrow."

Enrique turned to follow Captain Davis out of the office but was stopped by Agent Jones, who grabbed his elbow in a firm grip.

"Before you leave, Mister Lopez, can you tell me if you spoke with anyone in town? Any strangers?"

Enrique shook his arm free of the man's grip and shot him

an angry glare.

"You mean other than the spooks you sent?"

Agent Jones looked questioningly across the room at the other special operative, who shook his head. The agent frowned.

"Any further questions will have to wait until tomorrow," said the general, and Captain Davis escorted Enrique from the general's office.

"He's not telling us everything, General," said Agent Jones when the two had left.

"It doesn't make sense," replied the general. "By all accounts, Professor Appleby was his best friend on the base. Other than her and Captain Davis, Enrique doesn't really fraternize much."

"It's just too much of a coincidence," said Agent Jones. "Our launch of Project Juno is finally on track, and now the head of the project, the professor, goes missing. On that very same day our young Mister Lopez, whose father works alongside the professor on this project, decides to sneak off the base? I don't like coincidences, General."

"Could this be about Nephilim?" asked the general.

"Unlikely. Project Nephilim is the best-kept secret in the United States. Juno is an immediate threat to the Soviets and my instinct tells me that they've got their hands in this one."

"Nonetheless, Agent Jones, I suggest we increase the threat-level on Nephilim to Level 1."

After leaving the office Enrique followed Captain Davis across the base toward his living quarters. His mind was racing. Enrique knew Professor Appleby would never leave the base without telling anyone, which meant that something bad must have happened to her. He felt as if he were about to be sick. The best friend he had on the base, a woman who treated him like a son, was missing, possibly dead – and Agent Jones obviously felt that Enrique had something to do with the disappearance.

Who were these spooks, who came and went as they pleased? Who do they work for? The agents had asked about his father. Did they think Dad was involved? More importantly, who could have kidnapped the professor on a base with this kind of security?

The questions began to make Enrique's head throb. Everything was so confusing. But one thing was certain. His friend, Professor Appleby, was missing and he decided right then that he would get to the bottom of this mystery.

CHAPTER 4

Beyond the Garden of Homunculi

Dressed in faded denim overalls and a plaid flannel shirt, Stefan Gelemne walked down his dirt driveway to collect his mail. When he had first come to America several years ago, he had searched for a place with lots of room for crops. Finding the perfect location outside of Pahrump, Nevada, he had purchased two hundred acres, tilling and sowing them by day, and doing much, much more under the cover of darkness.

His driveway was so long that the house wasn't visible from the road, which was just the way Stefan liked it. Although friendly enough – many in town would have even called him charismatic, if they knew what the word meant – Stefan also enjoyed his privacy, and took somewhat severe measures to keep people away from his land. And if anyone ever wondered how one man managed to work so much acreage, or what he was doing out here, their questions were silenced when Stefan brought the plentiful fruits of his labor to market and sold to local stores at very reasonable prices. Stefan fit in well with the other local farmers; as long as he produced his record crops, his hard work spoke for itself, and if there was one thing folks in Pahrump respected, it was good, honest labor.

As he checked his weather-beaten mailbox, finding nothing inside but a battered postcard, a well-worn pickup truck rattled down the dirt road. Stefan exchanged a nod with the farmer behind the wheel, and stared at the dusty vehicle as it passed. He caught the eye of the farmer's wife as she glanced at him, and dropped her a sly wink. She hid a smile behind her hand and quickly looked straight ahead again. Stefan held his genial expression until the truck was out of sight, when his features twisted into an ugly sneer of contempt.

"It is like living among talking cattle out here," he mused as he flipped the postcard over to read its brief inscription:

Dear Stefan,

I look forward to our visit at the end of the month.

Yours,
Uncle Robert

Checking the postmark, Stefan shook his head in disgust. The card should have been delivered two days ago – so much for the U.S. Post Office. No matter, he thought. Everything would be ready for his "uncle's" visit. It always was.

He shaded his eyes with his hand and looked down the road at a distant plume of dust rising from another car approaching. Stefan decided to wait for his "uncle" and enjoy the ride back to the house in his fine motorcar. He broke a contemptuous smile as he thought of the strides in technology mankind had made over the past few decades, and his smile stretched even wider as he pondered his own contributions to the modern world – great and terrible contributions that would soon change it forever.

The car, a brand-new blue 1956 Packard Patrician, pulled to the side of the road and a mustached face leaned out of the open window. "It is good to see you again, nephew." Although his black hair and broad face with dark eyes below thick eyebrows revealed the driver's Eastern European heritage, his voice was pure Midwestern American, with no trace of an accent.

Stefan's "Uncle Robert" was in reality KGB Colonel Vitaly Delekarov. He was also Stefan's handler, in charge of overseeing his espionage activities, and visited the farm at least twice a month for progress reports.

Stefan walked to the other side of the car and got in, glancing at Delekarov as he sat down. The KGB colonel, a squat, stern-faced man, was dressed in a striped polo shirt with flannel pants and brown loafers. It was a casual look, popular with many American men of the day, and was designed to not attract attention. Hiding his contempt, Stefan nodded in greeting. "And you, Uncle, always a pleasure." Though his body was thin and frail-looking, his voice was surprisingly deep and rich. He seethed inwardly at having to address the younger man by the ridiculous appellation. Although Stefan appeared to be in his late forties, he was in reality almost a half-century older.

On June 30, 1907, the man who would eventually become Stefan Gelemne had awakened on the blasted ruin of Krasnoyarsk Krai, better known as the Tunguska Wilderness. His own countrymen had exiled him here, broken and beaten. Sentenced for crimes of unspeakable degeneracy and wickedness, he had been abandoned here to die.

But he was not like other men. Within him burned a black desire for revenge on those who had exiled him. This desire was fueled by unearthly abilities, including mesmerism, legerdemain and unnatural feats of healing; skills that assisted him in his quest for control over other men.

In the court of Tsar Nicholas II, Emperor of Russia, he had made his first bid for power. His abilities of healing and mesmerism endeared him to the Tsarina, and brought influence and reputation. But his ascendance as Gregori Yefimovich Rasputin, "The Mad Monk," was short-lived, and soon his enemies rose against him. Poisoned, shot and clubbed, Rasputin was thrown into the icy Neva River and left for dead. But as in the Russian wilderness, he would not die so easily.

In 1943, he rose to prominence again, this time in Germany as the infamous Dr. Josef Mengele. As the dreaded "Angel of Death," he indulged his appetite for evil and depravity, using new medical technology developed by the German state. But the Nazis were driven from power and their henchmen killed or fled into hiding. And the most hated henchman of them all, Dr. Joseph Mengele, would transform himself again, adopting the identity of Belgian immigrant Stefan Gelemne and settling on a farm outside the quiet town of Pahrump, Nevada.

But no matter what name he took or where he lived, Stefan harbored an insatiable lust for power and cruelty. These nefarious qualities made him one of the most valuable assets of the Soviet spy agency, the K.G.B., which had recruited the displaced Dr. Mengele in 1946 and established his cover as Stefan Gelemne in the United States.

As Colonel Delekarov turned into the driveway, he wasted no more time on pleasantries. "What is the status of our operation?"

"The mole has obtained a launch date for the American satellite, Juno," Stefan answered. "It is October first."

"That is unacceptable," Delekarov replied. "Sputnik is not scheduled to launch until the fourth of October. We cannot allow the Americans to beat us into space."

"I understand, Colonel. And they shall not. Our mole has initiated plans to ensure that Juno will never leave the launch pad."

"Excellent," said Delekarov.

The colonel pulled up to the farmhouse, and both men got out of the car. Stefan led him to an unassuming wooden toolshed behind the house. Inside, it looked like many rural toolsheds, with a work table, various disassembled machine parts scattered around, and a tool rack against the back wall.

Stefan went to the rack and pulled on the handle of a metal trowel. The work table slid across the floor, revealing a dark stairwell leading underground. He led Delekarov down the stairs, which ended in a broad, iron door. Taking a large key from his overalls, Stefan inserted it into the lock, pushing the heavy door to reveal a well-lit labyrinth of rooms and hallways that extended as far as the eye could see. It was like an expansive dungeon from the Dark Ages, updated for the twentieth century.

Stefan and Colonel Delekarov proceeded down the main passageway. On both sides were rooms where people were working on various espionage-related activities. In one room, attendants dressed in white lab coats mixed chemicals for poisons and explosives, in another room several workers constructed what looked to be futuristic weapons. In another, several workers were fabricating rubber masks and lifelike wigs. The end of the main passageway ended in another heavy iron door, which Stefan opened with the same key.

This room was at least one hundred yards long and seventy yards wide. Against both walls were giant banks of computers that clicked and whirred. Several technicians checked computer readouts and made notations. As Stefan and the colonel walked down the laboratory hall, they came upon two rows of what appeared to be giant glass test tubes. About the size of a wooden water barrel, each test tube was full of a percolating green liquid illuminated by an eerie incandescence. Inside each of the nearly two-dozen containers were the lifeless bodies of small, grotesque creatures. They were uniformly small, and although their heads were humanlike, their bodies included traits of various other creatures ranging from monkeys to spiders to ants. Each was about the size of a newborn baby and possessed wicked-looking teeth, claws, horns, or pincers.

Stefan and the colonel passed these test tubes without a glance. Occupying the rear of the laboratory was a large patch of soil approximately twenty yards across by twenty yards deep. The square was surrounded by wooden rails, and was lit by row after row of suspended, bright white lights. It looked like an indoor garden. But the good farmer wasn't tending any ordinary crop here. The soil was blood red, and among the furrowed rows of dirt there were at least a dozen of the creatures buried up to their necks. These, however, were alive.

Each of the creatures had a queerly humanoid face, except their mouths were much larger and contained rows of sharp pointed teeth. Their eyes were huge and glowed a sickly yellow. The noise these grotesque creatures made, a hellish cacophony of hideous screeches, filled the underground enclosure.

"I see your homunculi garden is proceeding as planned," said Delekarov.

"Indeed," said Stefan. "But a more pressing matter has come to my attention."

"What could be more important than your genetic research?" asked the colonel.

"The Americans have Nephilim in Area 51."

Delakarov's eyes widened. "What? Are you sure?"

"Of course," Stefan said, walking to a row of computer banks and inputting data. "It is the New Mexico object."

"Has it been activated?" the colonel asked.

"Of course not," Stefan said. "The Americans are far too ignorant to decipher the activation process."

"Do not underestimate our enemy," Delakarov replied. "The Nephilim technology must not fall into their hands."

Stefan shook his head. "And it shall not. I have created a sleeper that will infiltrate the base."

"A sleeper?" The colonel frowned. "An agent?"

"Of a sort. It is a new creation." Stefan pushed a large, red button on the computer console. The chamber echoed with the sound of metal grating against concrete. Beyond the homunculi garden, two giant iron doors swung open to reveal a large, mechanical, multi-limbed robot. As it stepped out, the

chamber shook under its tremendous weight. It was about the size of a pickup truck and moved, somewhat awkwardly, on four metal appendages. As it approached, the mechanical monstrosity's four metal arms swung back and forth, as if searching for something to grab.

"The sleeper's mission is to activate the Nephilim," Stefan said. "Once it is activated, I will remotely control the Nephilim and bring it here. Soon the technology will be ours. And if Groom Lake Base and its vaunted Area 51 suffer severe damage during the process, then so be it."

"You have outdone yourself again, Stefan," said the colonel. "But how do you expect that aberration to infiltrate Area 51 security?"

"Leave that to me," Stefan replied, staring up at his monstrous mechanical creation with pride.

CHAPTER 5

Unanswered Questions

Enrique had heard his father, Esteban, come home late the night before, well after midnight. And although he was wide awake, Enrique feigned sleep when his father checked his bedroom. He did the same when his father looked in on him before leaving for work shortly before sunrise.

Enrique had been tossing and turning the entire night. It was difficult for him to accept that Professor Appleby was missing. It didn't make sense. The professor was one of the friendliest, best-liked people on the base. Who would want to do her harm?

Enrique had been spending an increasing amount of time hanging out with Professor Appleby on the base. She was different from most people at the top secret facility, who tended to be rigid and soldierly. Professor Appleby had a free spirit ... an open mind. She was willing to embrace new and different things. In fact, it was Professor Appleby who introduced Enrique to "rock 'n' roll" music; music from the likes of Bill Haley and the Comets, Elvis Presley and Chuck Berry. No one else on the base, not even some of the other kids seemed to "get" the message of rock and roll. But Enrique had. He found solace in the upbeat tempos and exuberant rhythms. He liked the fact that most of the adults and kids on the base frowned upon the "jungle music" (as they called it). He liked to be different; and so did Professor Appleby. She even let Enrique keep some of her rock 'n' roll records so he could listen to them in his room.

Enrique thought about the last time he had seen the professor. She had given him a new record to hold onto but had made him promise that he would wait to play it until they could listen to it together. They were also supposed to go to the new James Dean and Sal Mineo movie, *Rebel Without a Cause.* This was something they had planned on for almost a month, but the professor had cancelled on the day of the show. Enrique had thought nothing of it at the time. His friend had given no indication that she was upset, fearful or unhappy. On the contrary, she seemed excited about her work at the laboratory and felt she was near an important breakthrough. Apart from being sincerely apologetic about having to cancel the engagement, she had acted entirely normal.

That was the last time he had seen her. It was later that

evening when Enrique decided to slip off base after his father declined to take him to see the show.

Unfortunately for Enrique, who was too young to attend the movie without an adult, he was caught trying to sneak into the theater. He probably would have been given a slap on the wrist and sent on his way, but Enrique didn't take to kindly to the theater owner's bullish attitude and had smarted off. It was the theater owner who then suggested a night in jail might teach Enrique a lesson and Sheriff Hank had been more than happy to oblige.

Although he knew deep down that he had nothing to do with Professor Appleby's disappearance, Enrique could not help but wonder if things would have been different if he had chosen not to sneak off the base. If he had insisted the professor take him to the movie, would she still have gone missing? Maybe this whole affair would never have happened if he had done that.

It took some effort for Enrique to convince himself to get out of bed. The last thing he wanted to do was go to classes today. He could only imagine what some of the other kids were saying. They already thought of him as an outsider. A Mexican. A Greaser. And now some people at the base, specifically Agent Jones and his detail of security goons, thought he – or even his father – had something to do with the disappearance of Professor Appleby. Well, if he was going to get to the bottom of Professor Appleby's death, he wasn't going to get anywhere lying in bed all day.

As Enrique prepared himself for school, he reviewed what he knew about the professor's disappearance. She had been working on something important, important enough that she'd break a promise to take Enrique to the movie.

The mysterious Agent Jones had mentioned something about a sub-laboratory 14, Alpha sector that was located in Area 51. He thought his father had mentioned something about sub-laboratory 14 or Alpha sector, but he wasn't certain. He and his father did not speak much, and his father especially did not talk much about his work at the base.

Enrique needed more information. He needed to know what the professor was working on. Was there someone on the base who held a secret grudge against her? Agent Jones appeared to believe that foul play was involved. Enrique could not imagine why anyone on the base would cause her harm. As far as Enrique could tell everyone had treated her like their grandmother. Did that mean an outsider was involved? But how would anyone without proper clearance get onto the base? Although security was not necessarily designed to keep people from leaving the base, it was most definitely designed to keep them out.

Enrique's father had left a note for him in the kitchen asking his son to wait for him after work. "I need to talk to you – very important," the note read. Enrique crumpled the piece of paper and was going to burn it, but when he reached for his new lighter he found that it was missing. He searched all the pockets of the jacket and then his pants ... but nothing.

That's strange, he thought. He distinctly remembered the guard giving him back the lighter. He made a mental note to check the jeep he rode back to base in and to retrace his steps from the command center. He felt obligated to hold onto the lighter so he could return it to the biker, Sonny, in case he ever saw him again.

CHAPTER 6

The Mole

"Mole calling Farmer, this is Mole calling the Farmer, can you read me? Over." The man who called himself Mole sat at a communication console that featured a number of toggle switches, knobs and dials.

He adjusted several dials and called again. His voice was melodic and sibilant and, although it was somewhat muffled due to the surgical mask, one could not mistake the excitement in the voice. The Mole sat in a darkened laboratory. Around him were the instruments and apparatus that one would find in a medical laboratory; specifically a laboratory used to conduct animal experiments. But there were no animals present in the room ... except for one; a hideous creature that hid somewhere in the shadows of the chamber.

The communication console made a whistling sound that echoed through the laboratory and then a deep baritone voice with the hint of an accent replied.

"This is Farmer. Why have you broken radio silence?"

"I apologize, Commander, but I have discovered the precise location of the Nephilim," said the Mole in a submissive, yet excited voice. "It is in Alpha sector sub-basement 20. This location is heavily guarded, nearly impenetrable."

"That will not be a concern," said the voice of the Farmer.

"But, sir, you must understand, the security around this project is like nothing I have ever seen."

"Your mission is to disable Project Juno, not to worry about Nephilim."

"Yes, and Juno shall destruct upon launch, but, but … Commander, is it true? Is Project Nephilim all that Professor Appleby says it is?" The Mole could barely contain his excitement.

"It is all you have heard and more."

"Unbelievable," said the Mole. "But then we must do something. We cannot let the Americans study it."

"That will not happen. A sleeper has been placed on the base. It shall deal with Nephilim. You have your orders, and you will obey!" The Farmer's voice was firm and imperious. "Do what you wish with the professor, but for now you shall not move against Nephilim! Do you understand, Agent Mole?"

"Yes, Commander. I understand … Mole out."

For a moment it was silent in the chamber; then the Mole flew into a rage, knocking a pile of books off the table and throwing a stack of files into the air. He wailed in desperate anguish, the eerie sound echoing throughout the chamber. And when the keening cry died away, a faint laugh came from the far side of the laboratory. It was a woman's laugh.

"Quiet!" shouted the Mole. "Don't you dare mock me!"

"How quickly the traitor becomes the betrayed," said the woman's voice. "I must admit, I love the irony. No one deserves it more than you."

The Mole snatched a large metal flashlight from atop the console and crossed the laboratory. He flipped the toggle on the flashlight and pointed it toward the ceiling. The flashlight revealed the laboratory had been built inside a cave. Large, pointed stalactites hung from the roof. In a corner of the chamber, looking very much like a fly caught in a spider's web, was Professor Appleby. She was wrapped up to her neck in a white, stringy cocoon, and the cocoon was affixed to the cave ceiling and walls by thick lines of webbing. Her hair was disheveled, her face bruised and pale, yet she had a defiant look in her eyes.

The Mole removed his surgical mask and glared dangerously up at her.

"Do not make the same mistake as the commander, dear Professor," said the Mole. "Do not underestimate me. The commander may have the sleeper but the Nephilim will be mine, I can assure you of that."

CHAPTER 7

Pecking Order

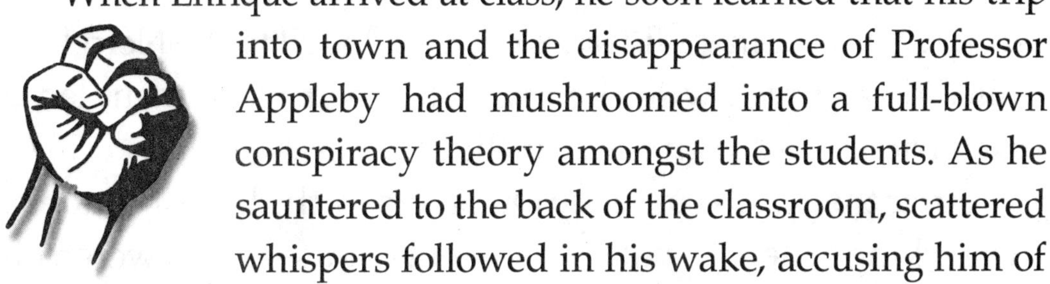

When Enrique arrived at class, he soon learned that his trip into town and the disappearance of Professor Appleby had mushroomed into a full-blown conspiracy theory amongst the students. As he sauntered to the back of the classroom, scattered whispers followed in his wake, accusing him of kidnapping, murder and even being the worst of the worst – a Communist. Enrique didn't bother to defend himself or correct anybody, he just slouched down in his seat and waited for the last bell of the day, keeping an eye on the rest of the kids around him as he tried to wait out another day.

Even at a top secret military base in the middle of the Nevada desert, kids are required to go to school. And just like most other schools, the Groom Lake Academy had its pecking order and cliques. At Groom Lake the top of the social pecking order belonged to the "Brats." Brats was a term of endearment for the school-age sons and daughters of the active military personnel on the base. The boys were mostly jocks. They dressed in preppy clothes and wore hair styles – crew cuts and flattops – that conformed to their parent's ideals. More than any other base he had been on, Enrique noticed that the Brats at Groom Lake carried themselves with an air of entitlement and superiority in relation to other kids. Enrique concluded that part of this was because of the high level of security and compartmentalization at Groom Lake the base. Because of the top secret designation there was a degree of seriousness that permeated all aspects of base life; this even extended to the cliques within the base academy.

After the Brats, next in the Groom Lake pecking order would be the sons and daughters of the scientists: "The Nerds." The scientists were highly valued on the base and received benefits and liberties that even some of the military officers did not receive. Some of these liberties even extended to the Nerd kids. This didn't sit well with the Brats and their resentment led to much friction between the two cliques.

After the Nerds came the Outsiders – the lowest clique on the pecking order. Usually the parents of these kids weren't in the military and weren't scientists. They were short-timers; workers brought in to do a specific job and then sent away. Outsiders seldom hung around long enough to find their way into either the Brat or the Nerd cliques. Enrique was an Outsider. But it was different for him. His father had found a valuable role at Groom Lake and had been there for over a year, much longer than short-time workers. Despite the length of time he had been at the base, Enrique never caught on with either the Brats or the Nerds. While the Nerds were tolerant, many of the Brats found it amusing to ridicule Enrique's Chicano heritage.

Several times during the day Enrique's teacher, Mister Pan, had to interrupt his lesson and warn the class about their whispered gossiping. But it wasn't until the lunch period that Enrique saw how blown out of proportion his misunderstood role in this affair had been become. At the bell, his teacher, Mister Pan, a sweaty, ill-tempered man, pulled Enrique aside.

"Mister Lopez, at the request of General Johansson and other authorities here at Groom Lake, I have given you a degree of latitude I would never give most students."

"When you say "most," do you mean white students?" asked Enrique in a cynical tone.

The vein on Mister Pan's temple began to throb.

"You know full well what I mean!" said Mister Pan. "You're not the first student who has lost a parent. You're not the first by a long shot. This is a military base, young man, and everyone who serves here, soldiers and civilians alike, knows the risks when they sign on. And while General Johansson and the others may believe your impudence and disrespect for authority are your ways of coping with grief, I see you for what you are: a selfish, ill-mannered and undisciplined lout! And I will not tolerate your unruly behavior influencing the other students in my class. Do I make myself clear?!"

Enrique defiantly held Mister Pan's stare, his mouth tightening into a thin line as the silence stretched out between them. "Don't you throw daggers at me, Mister Lopez!" said the teacher. "Do I make myself clear?"

Reluctantly, Enrique looked away. "Crystal," he said.

Enrique collected his books and exited the portable classroom. He could not think how things could get worse.

"Yo, Greaser!" The call came from Eric Duggan, a hard-looking acne-faced Brat tough who made it his business to bully other students ... especially Outsiders.

"Because of you Greaser, the rest of us have lost all liberties off base. And for that you're gonna get a beating."

CHAPTER 8

Duggan the Bully

Eric Duggan was far and away the largest student in Enrique's class. That was because he had been held back two grades. And Eric enjoyed the fact that he was older and bigger than the other kids.

Enrique had stopped at the top of the portable stairs. The rest of the students had gathered in a semicircle behind Duggan; Brats in the front, Nerds in the back, Outsiders on the perimeter. Enrique had seen this situation before and it usually ended in one of three ways: 1) Duggan gave a student a beating; 2) the student begged for mercy and had to kiss Duggan's feet; or 3) a passing adult intervened. Just two months previously, Duggan had given Enrique a shiner the size of a plum and Enrique still had to kiss Duggan's boot. It was after that incident that Enrique began taking Aikido lessons from Captain Davis. But any confidence he had in the Japanese martial art evaporated as he looked at the brutish Duggan and his mob of Brats.

Suddenly, the malevolent smile that Duggan had been wearing dropped from his face. His aggressive posture softened. From behind, Enrique heard Mister Pan step out of the portable doorway. Enrique let out a sigh of relief. Not even Duggan was stupid enough to provoke a fight with the teacher there.

Finally, thought Enrique. I catch a break.

Enrique started down the portable stairs, prepared to confidently walk right past Eric Duggan, staring him straight in the eye the whole time, but before he hit the bottom stair,

he heard the portable door close. Duggan's ape-like laugh told Enrique that Mister Pan had gone back inside and had no intention of coming out any time soon.

Before Enrique could turn back, the bully knocked the books from his hands. The rest of the class had already spread out and encircled the two combatants. Enrique desperately looked around for an avenue of retreat but there was none. Behind the smiling faces of the Brats, he spotted the sympathetic looks of the Nerds and Outsiders.

"Well, Greaser," said Duggan, "It looks as if this is my lucky day."

"Alright, Duggan, punch me in the stomach or the face and let's get this over with."

"No I don't think so. I think this requires something special."

"Why don't you leave him alone, Duggan? He's already in enough trouble as it is." The comment came from Carol Jean Johansson. She was a tall, gangly red-haired girl who was in Enrique's class. Although she was General Johansson's daughter, and should have been a natural fit with the Brat clique, Carol was a full blown, tried and true, geeky-as-can-be Nerd. Carol pushed her way to the front of the mob.

"Would you please not make this worse for me?" Enrique said to Carol.

"Yeah, Brace Face, why don't you make like a banana and split." Duggan let out another ape-like laugh.

"Now where was I ... " said Duggan. "Oh yeah, I want this to be memorable. Here's what I'm going to do. First I'm going to pound on you, then I'm going to take that leather jacket, then I'm going to drop a world record dribble-spit into your mouth and make you swallow. How does that sound?"

An unusual thing happened to Enrique just then. A moment ago he had resigned himself to receiving another humiliating beating at the hands of Erik Duggan. Sure he would put up a fight, but the beating would come all the same. There was nothing he could do about it. The coming pounding was just another in a never-ending series of hardships and ordeals that fate seemed determined to pile on him. Where it went wrong for Erik Duggan, however, was that the bully's taunt reminded Enrique of a piece of advice Captain Davis had given him during an Aikido lesson. "Never fight in anger ... anger makes one vulnerable." So, out of desperation, Enrique decided to stir the hornet's nest.

"Why do this to yourself, Duggan?" Enrique asked in an

even, confident voice. Duggan looked around as if he was missing something.

"What was that, Greaser?"

"I said, why do you do this to yourself? Why do you insist on making yourself out to be a bigger idiot than you already are? It's like every day you take a handful of stupid pills and come to class. Tomorrow, try laying off the stupid pills."

There was a moment of silence; then the other students broke into a chorus of laughter, especially the Nerds and Outsiders. There were even some hoots and hollers. Enrique could see the level of rage building in Duggan's neck and face. He inhaled a long drawn snort and with all his might expelled a hefty green loogie at Enrique. But the arcing trajectory of the slurry-like phlegm sent it high and wide. The loogie hit Priscilla Winthorp – who was standing behind Enrique – directly in the face. Priscilla was the unofficial queen of the Brats. Arguably the prettiest girl at the base school, she was definitely the most conceited, and was a budding love interest of Duggan's.

The bully's jaw dropped and the blood drained from his apple-red face when he saw where his spit had landed.

"EWWW!" Priscilla let out a horrified screech.

Enrique seized this as his moment to act. He quickly stepped toward Duggan and sharply thrust a knife-hand blow directly into the bully's throat. This was one of the first techniques Captain Davis had taught him. The captain explained it was very useful against much larger opponents, and now Enrique could see for himself how true that was. Duggan grabbed his throat and fell to his knees, desperately trying to catch his breath.

"Come on!" Carol Johansson grabbed Enrique's wrist as

she turned back and shouted to a short fat kid dressed in khaki shorts, white shirt and cowboy hat. "Tommy, grab Ric's books!"

Carol dragged Enrique through the crowd of students and walked him briskly toward the center of the base. The cries of Priscilla Winthorp could still be heard above the laughs and exclamations of the other students.

"Wait up!" The call came from the kid who was struggling to cradle at least a half dozen books as he ran, his cowboy hat bouncing precariously on his head. "Thanks," said Enrique as he took his books from Tommy.

"Smooth move, Ex-lax," said Carol sourly.

"Who? Me?" asked Enrique.

"Yeah, you," said Carol. "Now he's going to be really mad. You must have a death wish. Boy, I'd hate to be in your shoes."

Enrique halted.

"Well, you're not in my shoes. And I didn't ask for your help either."

"That's right, you didn't, but you need it, and so does Professor Appleby," said Carol.

"You know what's happened to Professor Appleby?" asked Enrique

"Not yet, but I'm hoping to find out."

"Then we need to see my father. The spooks think he's involved in all of this."

"The spooks?" asked Tommy with a tinge of fear in his voice.

"I don't think seeing your father is a good idea right now, Ric," said Carol.

"Why's that?" asked Enrique.

"Because the spooks have him in detention."

CHAPTER 9

Alpha Sector

Although Enrique was in the same class as Carol Johansson and Tommy Reilly, he had never really gotten to know them. Truth be told, he hadn't really gotten to know anybody in his class. Ever since his first day on the base when Erik Duggan and some of the other Brats had made disparaging remarks about his Latino heritage he hadn't felt like trying to get to know any of the other students. He always found that a few of the adults at the base were much more tolerant and understanding. That's why he enjoyed hanging out with Professor Appleby and Captain Davis.

Enrique had come to understand that while Priscilla Winthorp was considered the queen of the Brats, Carol Jean Johansson was regarded as the president of the Nerds, and not just because she spoke with a lisp and was the only kid on base who had a mouthful of shiny silver braces. It was also Carol's academic accomplishments and that she was the base commander's daughter, which set her several notches above the run-of-the-mill Nerds. Enrique also had heard that Carol was involved in extracurricular programs that actually allowed her to assist in some of the research that was being done on the base; or at least so it was said. Enrique always found her a little too bossy for his liking. A bit of a know-it-all, he thought.

Tommy Reilly was a different sort of Nerd altogether. While Carol was industrious and eager, Tommy was laid back and a bit of a slacker as far as Nerds went. A swell enough guy, Enrique thought.

Carol led Enrique and Tommy past the mess hall and toward the command center. As usual the sun was shining and the Nevada desert seemed to radiate the heat.

"So where are we going?" Enrique asked.

"We're going to Professor Appleby's lab," replied Carol.

"I hope it's not far," said Tommy as he labored to keep up with Carol's fast pace.

"Which building is it?" asked Enrique.

"You're telling me you don't know where your dad's lab is?"

"My dad's lab?"

"Yeah, Ric, your father worked with Professor Appleby; they were in charge of Project Juno. What, you didn't know that?"

"My father and I don't talk much," said Enrique, "I mean, I know they worked together, but whenever I asked him what he did, he said it was top secret."

"It is top secret," said Carol. "However, if you're part of the internship program like I am, then you get to help out with some of the projects." Carol's voice had taken on a somewhat haughty tone.

"I even have top secret clearance. I don't know if your dad told you, Ric, but I work with him and the professor on Project Juno," Carol continued, "Very top secret, Area 51 stuff. I've actually worked on a number of projects. I've seen places and things in Area 51 that you would never dream of, Ric. In some places my security clearance will let me bring friends. I've shown Tommy several projects ... "

"Yep, she has," said Tommy.

"Where we're going now, Ric, your dad's lab, it's in Area 51, but I can get you in."

"Would you stop it!" interrupted Enrique.

"Stop what?" asked Carol.

"My name is Enrique! En-rrrrreeee-kaaaay! Not Ric."

"Sheesh, relax," said Carol.

"Can we just get to wherever it is we're going? This 'top secret' lab?"

"Fine ... Ric."

About fifty yards south of the command center, a small dirt road cut east, away from the center of the base. Carol led them at a brisk pace down this road, which wound between several rocky outcroppings and through a patch of mesquite trees.

"Ugh," said Tommy as he labored to keep up with Carol's fast pace. "How much farther is it?"

"We're almost there," said Carol.

Carol led them past a security checkpoint. The sign at the gate read, "Area 51, Security Clearance Required." They continued down a road that took two more turns and then ended against the side of a rocky bluff.

"Well, here we are."

Built against the bluff was a small, cube-shaped building made of thick blocks. The building was no more than twenty yards long by ten yards wide and was surrounded by two separate barbed wire fences. There was a full contingent of guards on duty.

"Whoa," said Tommy. "This is Alpha sector."

"Relax," said Carol. "I have clearance and you're with me."

Carol led Enrique and Tommy to the gate of the first fence line. The guard contingent all remained positioned behind the second fence line. One of the guards called out.

"Hi, Miss Johansson. Swipe your security card and enter the pass code, please. Once the gate is open, your group shall promptly proceed to the second gate post and present your identification to me. Do not, I repeat do not, stray from the walkway."

Carol swiped her card and entered what seemed to be a ten digit pass code.

"Stay close," said Carol. "And stay on the path."

"Why's that?" asked Tommy.

"Because if you don't, they'll shoot you."

Tommy, noticing that all the guards did indeed have their rifles ready, scurried up close to Carol and walked in step with her.

At the next gate Carol repeated the procedure; swiping her security card, then entering a pass code. Enrique noticed that the pass code for this second gate was different than the first.

"Afternoon, Miss Johansson," said the guard.

"Afternoon, Sergeant Mullins. They're with me today," said Carol, gesturing toward Tommy and Enrique.

"I just need to see your base identification cards, gentlemen," said the guard to the boys.

"Good to see you're getting some help on your research project," said the guard to Carol.

"Yeah, if you can call it that," Carol said under her breath.

After Enrique and Tommy showed the guard their base identification, the guard escorted them inside the building. The doorway opened into a single hallway painted in olive green. The hallway was narrow and ended in a single elevator door about twenty paces from the entrance. Using a key, the guard opened the elevator. As they were about to enter, the guard stuck out his hand.

"Code seven point zero one four of section nine of the Area 51 Security rules and regulations prohibits the use of recording or photographic equipment inside any and all buildings ... "

"Thank you, Sergeant Mullins," interrupted Carol, "but we are all very familiar with the rules and regulations at Groom Lake. I think I can handle it from here."

"Roger, Miss Johansson. Just don't stray."

Enrique, Tommy and Carol proceeded into the elevator. The guard remained in the hallway and gave them a friendly wave.

"Are all the guards this nice to you?" asked Enrique.

"They are when you're nice to them," said Carol.

The elevator doors closed and Carol once again swiped her security card and punched another pass code into a console panel.

The inside of the elevator was so large it could easily hold two jeeps. Hanging on hooks against one wall were a half dozen white lab coats.

"Grab a lab coat, boys, and clip your identification to the pocket so everyone can see it," said Carol.

Enrique gave Carol a disapproving look and flipped up the collar of his leather jacket as if to say, This ain't coming off.

"Suit yourself, Ric," said Carol.

"Enrique!"

"Whatever. No lab coat, no entrance. It's the rules."

Reluctantly, Enrique removed his leather jacket and put on a lab coat. He folded his leather jacket and draped it over one arm.

Enrique felt the elevator slow, and then it seemed as if

it began to move horizontally. But Enrique knew that to be impossible. Elevators were suspended by vertical cables. How could they move sideways?

"Hey, check it out," said Tommy. "I'm a scientist."

Tommy had found a pair of safety goggles in his lab coat pocket and put them on. Enrique could not help but laugh. The goggles made Tommy's eyes appear enormous, which in turn made the rest of his features appear tiny.

"You guys better get serious," said Carol.

"Yeah, Tommy, you better get serious," said Enrique in his best Carol impersonation, lisp and all. Carol shot him a menacing glance.

"You know, I don't get it," said Carol as she faced Enrique with her hands on her hips."

"What don't you get?" challenged Enrique.

"In class, everyone thinks you're a sour apple ... " began Carol.

"I don't," Tommy said, wrestling the goggles off his head.

" ... but while I was working with your father and the professor, they had nothing but good things to say about you. And then in class today, when you stood up to that bully I thought there was something more to you than what the other kids say. But, you know what? I can see now I was wrong. You are nothing but a sour apple."

Before Enrique could respond, the elevator stopped and the doors opened.

"Whoa," exclaimed Tommy.

The elevator had opened onto an enormous, modern foyer made of stainless steel and granite and bustling with activity. The foyer was huge, with a number of different hallways intersecting and leading in different directions. The ceiling was almost as large as the airplane hangers on the base. Scientists, soldiers and engineers were busily going about their daily routines. Some pushed or pulled exotic looking machines, while others rode about in miniature carts that hummed along at a brisk clip. Two soldiers bounced past them on kangaroo-leg-looking pogo sticks. Several individuals actually flew overhead in loud jet packs. Another scientist raced past them wearing wheels on his shoes that shot flames from the back.

"Welcome to Alpha sector," said Carol.

CHAPTER 10

Professor Appleby Leaves a Clue

Carol briskly led them out of the elevator and through the bustling foyer.

"Try to keep up," she said. "And look like you belong here."

Carol led Tommy and Enrique down one of the large halls off the main foyer. Along this hall were a number of offices and laboratories. Some had signs that read, "Danger: Radioactive," while others had red warning lights that read, "Caution: Experiment in Progress." Some doors to the offices and labs were open and Enrique caught glimpses of test tubes, large computer banks and mysterious-looking machines.

Every now and then Enrique could hear Tommy behind him exclaim, "Whoa," or "What's that?" or "What does that do?" and after each comment Carol would admonish him with a "shush!" Every fifty feet or so the three teens would come upon a televisionlike monitor that was bolted to the wall above their heads. The monitor showed a real-time image of the hallway.

"Hey look, we're on TV!" exclaimed Tommy, waving.

"Knock it off, you guys," admonished Carol.

When Enrique was in the picture of the next television monitor he walked directly behind Carol, put his thumb to his nose, stuck out his tongue and wiggled his fingers. Tommy covered his mouth with his hands to stifle his laugh.

Eventually Carol stopped at a door that read, "Alpha Lab S-

14. Authorized Personnel Only." It had a yellow security tape fastened across the door.

NO ENTRY WITHOUT AUTHORIZATION
BY ORDER OF THE DEPARTMENT OF COVERT SECURITY

"Here goes nothing." Carol glanced around before inserting her identification card into the security console and punching a code into the keypad. With an audible click the door popped open. The three students ducked under the security tape and stepped into the room.

Carol ushered the two boys in, keeping an eye on the deserted hallway. "Let's get in and get out quick – the spooks might be keeping an eye on the lab." For once, Enrique agreed with her.

The laboratory was as futuristic as anything Enrique had seen in any science fiction magazine or movie serial. A large computer bank sat against the wall, still whirring and clicking with its two large spools of tape spinning around. Several tables held large microscopes and various tools and devices. Against another wall was a large blackboard covered with scrawled mathematical equations.

"What happened here?" Enrique asked, more to himself than the others.

Signs of a struggle were scattered throughout the laboratory; a knocked-over table, files strewn across the floor, a broken beaker. On the floor next to a file cabinet there was a crushed rack of test tubes and what looked like drops of dried blood. In the middle of the room was a smashed metal globe. The globe had various antennae and other appendages sprouting from it, many of them broken.

"What's this?" Tommy pointed at the globe.

"That was a model of Project Juno," said Carol. "Juno is a man-made satellite. It's the priority research program currently in development on the base. The spooks think the professor's disappearance had something to do with Project Juno."

"My dad was working on a satellite?" asked Enrique.

"Yes, he worked on its navigation system. What did you think he did?"

"I thought he was an engineer, you know, working on engines and stuff."

"He does, but it's for the launch rocket. I helped him calculate some numbers for the payload ratios. He is extremely knowledgeable, and very helpful and polite. Not at all like you, go figure. I guess sour apples fall farther from the tree."

"Whatever," said Enrique, glancing around at the debris as if he stared at it hard enough, he could solve the mystery right here and now.

For the next thirty minutes the three rummaged through file cabinets, desk drawers and boxes of research notes; all with no apparent luck.

Against one wall were two lockers. Taped on one locker was the name, "Esteban Lopez," and over the other, "Gloria Appleby." Enrique took out his pocket knife and picked the lock on his father's locker.

"Learn that in Sunday school, did you?" asked Carol snidely.

"Actually in a book about Houdini's escape secrets," replied Enrique, "I ordered it from the back of a comic book."

"Figures," said Carol, somewhat sarcastically.

Taped to the inside of the locker door was a picture of

Enrique's mom and another of Enrique with his dad. Enrique recognized the pictures immediately. They had been taken at Newport Beach. Enrique's father had taught him how to body surf that day. His mother had packed a picnic. It was one of his favorite memories.

"She's pretty," said Carol, coming up behind Enrique. Enrique slammed the locker shut and tried the handle to Professor Appleby's locker.

"Hey, that's the professor's personal stuff," protested Carol.

"We're looking for clues, aren't we?"

The professor's locker didn't hold much. Some articles of clothing, a few books and some records. Enrique flipped through the records, finding Elvis Presley, Jerry Lee Lewis, Buddy Holly and the Crickets and several others.

"I don't know how anyone can listen to that noise," said Carol.

"Hey, look over here," said Tommy.

Carol crossed to where Tommy was pointing at the ground. Next to the broken glass there were several tiny footprints. The footprints were circled in red chalk.

"What do you make of that?" asked Carol.

"I don't know, I've never seen a footprint like that," said Tommy. "And I used to hunt with my dad and Uncle Ernie all the time back in Texas."

"And look over here ... " said Tommy, " ... against the edge of the file cabinet, tufts of hair. Feel how coarse it is?" Tommy handed the hair to Carol.

"My dad mentioned that the investigators thought an animal had gotten loose here. Maybe we should ask Captain Davis about this, he's an animal expert."

"That's weird," said Enrique. "This record cover has a date

followed by the initials P.J. written on it."

"What's so weird about that?" asked Carol.

"Professor Appleby's a collector; she would never write on an album cover."

"Lets play it," suggested Tommy.

Next to the large computer bank was a small record player sitting on the table. Enrique carefully put the record on. The inevitable hiss of static projected from the record player and then a voice began speaking … it was Professor Appleby.

"Journal entry date March sixteenth, nineteen fifty six. Today we began work on Project Juno. I am very excited about this project. There is much work ahead of us but the scientific implications of this project are enormous … " There was a long pause in the record and then the voice began again.

"Journal entry date March eighteenth, nineteen fifty six … "

"It's a personal journal," said Carol. "All the scientists are required to keep them but I've never heard of a recorded journal before. Most are written. The initials P.J. must stand for Project Juno."

"Maybe she kept two journals, one written, and one recorded," suggested Enrique as he pulled the needle off the record and set it aside.

"Wait," said Carol. "This is what we're looking for."

"I know, but check the date; we want her last journal entry, not her first," Enrique replied.

He continued to flip through the records, pulling out the last one, a Chuck Berry album. Enrique read the writing on the album cover.

"August eighth, nineteen fifty six, then the initials P.N."

He placed the record on the turntable and set the needle. After a moment of static the professor's voice began.

"Journal entry date September sixth, nineteen fifty six. Despite my objections, General Johansson has turned over complete authority of Project Nephilim to Agent Jones and his thugs from the Special Operations Agency. I think this is a tremendous mistake. I do not trust these spooks, especially Agent Jones. My fear is that they will corrupt the peaceful and scholarly parameters of this project. I hope the general knows what he's doing."

The professor's voice ended and then there was static.

"That's the last track on the record," said Enrique.

"And there are no more records in the locker," said Tommy.

"Wait!" shouted Enrique. "The day the professor went missing she gave me a record. I didn't think anything of it at the time but she made me promise that I would wait and listen to it with her."

"What's Project Nephilim?" asked Tommy.

"I've never heard of it," Carol answered self-importantly, "and I've heard of most of the projects on the base, even the top secret ones; because of my top secret security clearance and all."

Enrique ignored her, ran to his books and pulled out a forty-five record that was stashed in between the pages of one of them. He held it up for Tommy and Carol. On its cover was written, "September 26, 1957."

"That was just a couple days ago!" exclaimed Carol.

Carol and Tommy gathered close as Enrique slid the record on the turntable and dropped the needle. This record had more white noise than the others. When the professor spoke, she had an edge of concern in her voice.

"Journal entry September twenty six, nineteen fifty six. My concerns have proven correct. There is an unauthorized broadcast originating from the base. It is in code but I have been able to decipher segments of the transmission. Project Nephilim is the target. A traitor has infiltrated Area 51. I don't know who I can trust. If the traitor were to activate the Nephilim object, the entire base could be in danger. I plan on meeting General Johansson later today in private. Tonight I look to confirm my theory regarding the harmonics ... "

Suddenly a loud squawk issued from the laboratory entrance. All three students jumped. Enrique quickly pulled the needle off the record.

"Sergeant Mullins calling Alpha Lab S-14. Are you there, Miss Johansson?" The voice had come over the intercom. Carol crossed to the intercom station and pressed a button.

"Yes Sergeant, what can I do for you?"

"The general is here and he wants to see you and Mister Reilly immediately," replied the sergeant over the intercom.

"Tell Daddy I'll be right there," said Carol.

"Immediately, Miss Johansson."

"Roger, Sergeant Mullins." Carol turned to Enrique. "Finish the record and then meet us upstairs. You know how to get back, right? How cool is this? We're in a mystery. I tell you, I know just about every secret there is on this base, but this Project Nephilim must be a doozy because I've never heard of it."

"What does she mean, the base is in danger?" asked Tommy.

"Come on, Tommy," Carol said, grabbing him and practically skipping out of the laboratory. To Enrique it seemed that she was actually enjoying herself. The door closed and Enrique turned back to the record player.

Outside, Tommy had to lift his lab coat so he wouldn't trip over it in order to keep up with Carol as she hurried down the hallway. The elevator door opened just as they arrived. Agent Jones stepped from the elevator and grabbed Tommy Reilly's arm. General Johansson was behind him and he reached out to his daughter.

"Are you okay, sweetie?" the general asked his daughter.

"Yes Daddy, what's wrong?"

"Enrique Lopez is what's wrong," replied the general. "We believe he is a traitor."

CHAPTER 11

The Sleeper Awakens

Enrique was being escorted down the hallway of sub-level 15 of Alpha sector by two armed, military guards. After he had listened to the startling remainder of the record, he had quickly left the professor's lab and run, literally, into Agent Jones and a security detail. With no explanation, they had handcuffed him and led him through the main foyer, down a nondescript hallway and then through a doorway guarded by two spooks and into a large clinical room. At the far side of the room there was an enclosure made, for the most part, of glass. The enclosure had four entranceways that Enrique could see. Each of the entranceways was lined with metal frames. Enrique had seen a laboratory like this once before, when he toured a quarantine facility on another base.

As Agent Jones and the guards led Enrique to the nearest entranceway, the glass door section, with a mechanical hiss, slid open as if by magic.

The laboratory inside was as sterile as a hospital room and contained diagnostic equipment and a number of computer banks. In the middle of the lab was a large table containing various tools and instruments. Against the far wall was a bank of translucent plastic panels – the kind that are used to view X-rays – and several large, boxy computers. On a table next to the computers were test tubes and beakers filled with different colored fluids.

Waiting for Enrique and Agent Jones in the laboratory were General Johansson, Carol, Tommy and Captain Davis. All were sullen and downcast. Captain Davis especially seemed disappointed.

Beyond them was a row of television monitors. On one was the image of Enrique's father. He was sitting in a chair with his hands together on the table. They were handcuffed! On each side of him were special operations agents. His dad appeared distressed and disheveled. One of the spooks was leaning close and appeared to be shouting at him, asking questions, but because there was no audio, Enrique couldn't hear what they were saying. He called out to his father and waved his arms but to no avail.

"What'd I do?" demanded Enrique. "What's going on?"

"We were hoping you could tell us," replied the general grimly, as he motioned to the television monitor that displayed Enrique's father.

"We're showing you this in hopes that it will convince you how serous this is, and how important it is for you to be entirely forthright."

"Carol, what's going on?" asked Enrique. Carol tried giving him a sympathetic look, but ended up turning away.

Agent Jones went to the surgical table and lifted up a white cloth. On the table was the metal butane lighter that Enrique had found in the pocket of his leather jacket.

"This 'device' was found on your father during a routine security check on his way into work this morning," said the Spook. "We understand that a security guard at post Bravo Zebra observed this very same device on your person when you returned to the base yesterday."

"Device? What's the big deal?" said Tommy. "It's a lighter."

Agent Jones shot Tommy a menacing stare that caused the boy to wilt behind Carol.

"It is not just a lighter," said General Johansson. "This device features technology never seen before." The general crossed to a bank of X-ray panels that held a number of different films. Each of the exposures appeared to be of the device.

"Its casing has proven resistant to all our attempts at accessing its inner components. Judging by our X-ray examinations, though, its transistor technology is way ahead of anything we've got. It is our assumption that it is a weapon of some kind."

"What is its purpose, Mister Lopez?" Agent Jones demanded.

"I never saw that thing before the other day," said Enrique. "And you know what? The more I think about it, the more I think one of your agents planted it on me outside the jail."

"What agents?" asked the spook.

"Is this a weapon, Mister Lopez?" asked the general.

"Look," said Enrique, "I think you need to listen to Professor Appleby's recording … "

"Enrique, why don't you tell them where you got this lighter from so we can clear this whole mess up?" asked Captain Davis.

"I swear, I thought it was just a lighter … " Enrique

stammered.

He looked more closely at the device. It appeared to be an ordinary lighter, aside from the fact that the scientists had secured it to the table by several thin steel wires and pins.

"Who gave you the device, Mister Lopez?" Agent Jones continued, his patience obviously running thin. "Is this a weapon? Who is your father working for?"

After a moment Enrique said, "It's looking the wrong way."

"What was that?" asked the general.

"The eagle ... embossed on the lighter. Its head is facing the wrong way. Yesterday it was facing toward the hinge."

Enrique bent forward to get a better look. Suddenly, the metal embossed eyes of the eagle opened. Enrique jumped back as the device began to vibrate and trill upon the table.

"Stand back, everyone!" shouted Agent Jones.

Several loud clicks and popping sounds came from the device followed by a loud piercing scream.

"What did you do, Mister Lopez?" shouted the spook above the din.

"I didn't do anything!" Enrique answered.

Suddenly, two thin, wire-like protuberances thrust out from the lighter. On the ends of the wires were two pinhead-size lights that steadily grew larger as the device pushed against its restraints. Another series of loud "pops" resulted in four leg-like appendages sprouting from the lighter. The appendages pushed against the table, breaking the wire restraints. The device then stood on its four legs and scuttled about the surgical table like some mechanical crab, its two glowing red orbs swinging like roving eyes. A series of increasing shudders culminated in another loud "pop" as two mandible-like appendages sprang forth from below the eyes.

Two armed soldiers came running into the lab. Agent Jones pointed to Enrique and shouted to the guards, "Arrest that man!"

"Clear the lab!" ordered General Johansson.

As the two security guards went to seize Enrique, the mechanical crab reared back on its hind legs and let out a screech. This stopped the soldiers dead in their tracks. The mechanical crab stood on its hind legs, its small appendages and metal dactyls waving slowly. Its bulbous eyes began to glow red and the steady trilling noise it made got louder. One of the guards brought up the butt of his rifle and delivered a blow to the creature, which tumbled end over end several times before righting itself. The trilling noise increased and the glowing orb-like eyes grew brighter. Both guards shouldered their rifles, but before they could fire their weapons, two bright scarlet beams of light shot from the robotic creature's eyes. The light enveloped both security guards, who had only a moment to let out stifled screams before their bodies disintegrated into a cloud of ash. Their smoking guns fell to the floor.

"Secure the lab!" shouted General Johansson. "We have a potential e-tech security breach!"

Alarms and horns sounded. The entire facility was enveloped in a red light. Metal gratings began dropping to sheath the laboratory windows.

"This way!" shouted General Johansson, "Carol, hurry!"

Heavy iron doors began to seal one exit after the other. Carol grabbed Tommy and went to exit through one of the doorways but the metal grating slid shut a moment too soon.

General Johansson instinctively started to retrieve his daughter, but before he could Agent Jones reached out and jerked the general out of the lab just as the last heavy iron door slammed shut, sealing the laboratory.

The mechanical crab now leapt from the table and scuttled about the floor. Its two bulbous, antennae-eyes fixed themselves on a complex-looking diagnostic machine in the middle of the room. The constant mechanical trill increased in pitch as if the thing was excited. Carol and Tommy were huddled together on one side of the room; Enrique was on the other, the mechanical creature between them. They were the only people left in the lab.

"Don't move!" shouted Enrique.

Using its metal mandibles, dactyls and other appendages, the robotic creature – working at incredible speed – began to systematically dismantle the diagnostic machine.

"What's it doing?" asked Carol.

"How should I know?" replied Enrique.

"You brought this thing onto the base, you traitor!" shouted Carol.

"No I didn't! I mean, yes I did, but I didn't mean to. Carol, you've got to listen to the professor's record. It's not me! It's not my fault!"

On the monitors next to Carol, the general was waving into the camera, trying to get their attention.

"Carol, look!" shouted Enrique pointing to the monitor. The general had a telephone receiver in his hands and was pointing to the panel.

"Punch the audio button," said Enrique.

Carol did so. There was a short screech and then General Johansson's voice came over a speaker.

"Carol, are you okay?" asked the general in an alarmed voice.

Carol and Tommy were pressed up against the wall, each

holding onto the other, both petrified. Their eyes were fixed on the mechanical crab as it ravaged the diagnostic machine.

"We're okay, Daddy," said Carol in a shaky voice. "But what is this thing?"

"It's a robot of some sort, an automaton," replied the general. "What is it doing?"

"It's tearing apart the diagnostic machine," said Carol. "It's, it's ripping out wires and stuff."

"It's doing more than that, General," said Enrique. "It seems to be cannibalizing the technology and appropriating it for its own use."

Sure enough, the robot was simultaneously dismantling the diagnostic machine and integrating the salvaged technology into its own system, all at a frenetic pace. Metal sheets were being welded onto its body with its laser eye beams, its appendages and mandibles were securing radio tubes, installing transistors and splicing wires, its proportions growing at an alarming rate.

"Why is it doing that, Mister Lopez?" demanded the general.

"I don't know!" pleaded Enrique.

"You do know, Mister Lopez!" shouted Agent Jones through the speaker, "And you better start talking!"

"Look, General," said Enrique "I don't care what you think, but we've got to do something quick. This thing is growing."

In fact, the additions and modifications made by the automaton had increased its size to that of a large rat. It had now finished cannibalizing the diagnostic machine and had moved to foraging among the tools, equipment and other items that had been spilt upon the laboratory floor.

The automaton picked up several scraps and held them up to its bulbous eyes as if evaluating their worth, then after a moment it threw the scraps to the ground in apparent frustration. Finally, it turned toward Tommy and Carol, who were huddled next to a bank of computers. The automaton scuttled toward them, its mandibles and pinchers reaching out, clacking together.

"Daddy, help!" screamed Carol.

CHAPTER 12

The Secret

The mechanical crab leapt in the air. Both Tommy and Carol let out bloodcurdling screams. But the device wasn't jumping at them. Instead, it landed on the computer next to them and began ripping it apart.

Carol and Tommy quickly moved next to Enrique.

"Quick, Carol," Enrique pleaded, "take my knife out of my pocket. I need to get these handcuffs off."

"Don't do it, Miss Johansson." It was Agent Jones' voice over the intercom. Carol froze, not knowing what to do.

"Carol please, I didn't have anything to do with this," said Enrique.

"Let's try these," said Tommy, who had reached into the pile of ash at their feet and pulled out the soldier's keys. Soon, Tommy had unfastened Enrique's handcuffs.

"Carol, listen to me," said General Johansson in a calm, yet desperate voice. "We're having difficulty opening the security doors from out here. Your identification card should work from the inside. Security code is seven-seven-one-nine-six-one. Go to the nearest access door, but move slowly!"

Carol, with Tommy fearfully clutching her shoulder, slowly took several steps toward the nearest door. For a moment the robot ceased its fury of amalgamation. The computer was in pieces on the floor and the robot was now the size of a hound. Its bulb-tipped antennae eyes searched the room as if to see what else it could consume, but nothing seemed to appeal to its appetite. It shuddered and made a sound like a mechanical belch. The constant electronic trill now had a deeper, more ominous tone to it. The robot, which now featured a number of dangerous and lethal-looking appendages, scuttled several feet toward Carol and Tommy.

"Ric!" Carol shouted.

Enrique took his leather jacket, which had been placed on a table by Agent Jones, and swung it sharply at the automaton, whacking it on its back side and sending it spinning. The robot quickly righted itself and turned on Enrique. A low, menacing, mechanical growl issued from somewhere deep inside it.

"Now!" said Enrique.

Carol and Tommy quickly darted to the door, Carol fumbling for her identification card.

Enrique swung his coat again and lashed out at the advancing automaton. This time, the robot grabbed the coat with one of its pincer appendages and yanked it from Enrique's grasp. Using its pincers, the automaton tore off one sleeve of the jacket and then the other. The robot then threw the leather coat across the room and advanced on Enrique, mandibles and pincers clacking.

"Um, guys!" shouted Enrique as he backed against the wall.

The robot's eyes were now glowing brightly. The trilling

noise it made had reached a deafening pitch. Carol turned, and saw that the mechanical creature had Enrique pinned against the wall. She looked frantically for a weapon, but finding none within arm's reach, grabbed a beaker full of soupy, greenish liquid and threw it at the robot. The beaker crashed against the mechanical creature. Turning, it shook its body like a dog after a swim as sparks began to erupt from where there were creases and joints in its plating. The automaton growled mechanically at Carol and then awkwardly reared back on its hind legs and spread its mandibles wide.

The robot leapt, but just as it did, Tommy pushed Carol out of the way. The automaton landed on Tommy's chest and dug its two pincers into his shoulders. Tommy let out a cry and fell to the floor. The robot straddled the boy's chest and brought forward a single metal dactyl. A sharp, translucent, needle-like protuberance emerged from the finger, and in one swift, decisive movement the needle was thrust between Tommy's eyes, into his forehead.

Carol screamed as a bluish liquid seeped through the translucent needle and into Tommy's forehead. Tommy's eyes rolled into the back of his head and his body went limp.

"No!" Carol shouted. She jumped to her feet and grabbed a chair.

"You little Frankenstein freak, that was my friend!"

With an overhead swing, Carol brought the chair down with all her might on top of the automaton, which flattened against the floor, its multiple limbs spreading out. She lifted the chair over her head and brought it down again with all her might and rage, but the robot, catching a chair leg with one of its pincers, began chewing up the leg with its pincers and mandibles. Wood flew as the mechanical creature chewed its way toward Carol. Dropping the chair and turning to run, the girl lost her footing and fell to the floor. She turned, expecting to see the robot ready to pounce, but instead saw it lying upon its belly, its legs flailing. Beneath the robot there was a puddle of liquid from the broken beaker. The automaton couldn't find traction on the slippery floor! It struggled to get its multiple legs beneath its chassis. Its mandibles and pincers clicked desperately as its legs stuttered, slipped and skated in a blur of futility upon the slick surface.

The robot finally extended one of its pincer arms and snagged the hem of Carol's lab coat. Its legs still spinning on the floor, the creature used the lab coat to pull itself closer to Carol. As she let out a sharp scream, the mechanical creature responded with a laugh-like shriek of its own, which turned to a snarl. Just as the automaton was about to seize one of Carol's legs, the lab coat ripped, allowing her to scramble a few feet away. Still having difficulty finding its footing, the automaton stopped its flailing for a moment and its bulbous eyes began to glow, its constant hum increasing in volume.

"It's going to use its lasers!" Carol shouted.

But just then Enrique leapt to his feet and charged into the surgical table, knocking it over between Carol and the mechanical creature.

The robot fired its death rays at Carol but one beam struck the top of the metallic table and ricocheted to the ceiling; the other ray beam bounced off the surgical table into one of its own antennae-like eye bulbs, melting it instantly. The creature had lost one of its death rays! Releasing a mechanical wail, it turned on Enrique. But just then, one of the security doors slid open behind the robot.

"Who opened that door?!" shouted General Johansson

from the intercom. "That door leads into the sublevels. Secure that door immediately!"

From beyond the doorway, there came a high-pitched whistle. The robot turned toward the door, then back toward Enrique. The whistle sounded again and, giving Enrique a final mechanical snarl, the machine scurried out the door and down the hall.

Carol, weeping softly, crawled over to Tommy, who was lying still on the ground. His face was bloated and had a

purplish hue to it. He was sweating profusely and moaning.

"Tommy ... " whimpered Carol. "Oh, Tommy."

Enrique grabbed Carol by the shoulder.

"Carol, I think the base is in danger," he said in a desperate voice. "That thing is after Project Nephilim. Professor Appleby had deciphered transmissions coming from inside the base. There's a spy, and the professor believed that whatever this Nephilim is, it can destroy the base. I think the robot is going to activate it."

"Then go, Ric!" Carol said between her sobs. "Get that thing."

Enrique turned and flew after the mechanical creature.

General Johansson's voice came over the intercom again, "Get me access to that lab immediately, and I want a location on that ... that thing."

Running down the security corridor toward the elevator, Enrique found it easy to follow the path of destruction that the automaton left behind. Every twenty feet or so he'd come across a communication panel, security door or computer station that was torn apart, its wires still sparking and crackling. Fortunately, the elevator was still working and when the elevator door opened on sub-level 20, Enrique heard – and felt – a quick succession of explosions. It sounded as if fireworks were going off.

He carefully made his way down the deserted corridor. The concussions had ceased, but now he could hear movement down the hallway, and could smell gun powder.

As he rounded a corner, Enrique came upon an eerie sight. A large vault-like door, easily three feet thick, stood open. Scorch marks pocked the walls and floor of the hall and there were several piles of smoldering gray ash scattered across the floor; obviously the remains of unfortunate soldiers. Two other soldiers, lying in the cavernous entrance of the vault, were either unconscious or dead. Enrique could hear movement inside the vault. Beside the entrance, the security console panel had been ripped out, its wires still smoldering. The warning light above the console was flashing but the alarm was silent.

Enrique carefully crept into the vault. The entranceway opened onto a small foyer. Offices were on the left, a stairwell to his right. As silently as possible, Enrique tiptoed down the stairs. Several times the stairwell curved, before it finally opened into a large room at the bottom where Enrique beheld a most incredible sight. In the center of the room, surrounded by computers, sensors and other diagnostic equipment, was a genuine flying saucer! It was the size of a tank, and featured a number of peculiar angles and curves. Enrique did not know why, but the thought that the disk may have been constructed by the hands of man never entered his mind. This was not an earthly craft. Human beings could not create something like the machine that stood before him.

Beneath the saucer was the robot; now the size of a horse, it lay twitching upon the ground. Sparks popped and smoke drifted out of a large hole in its battered chassis. The robot emitted one last desperate whine and then its appendages stopped twitching. Enrique gasped as he noticed – beside the automaton – a hospital gurney and upon the gurney lay – yes, an alien creature! A number of tubes and wires, including intravenous lines, led from the alien to a multitude of machines, bags and other equipment.

Fearfully, Enrique crossed to the gurney, having to step over several piles of ash along the way. As he approached, he could see the alien's chest rise and fall. It was alive! At the foot of the gurney, in the grasp of one of the robot's pincers, was an unfortunate dead soldier, still clutching his pistol. As Enrique looked down at the alien, he could not help but feel a kinship with the creature. Although it had an oxygen mask placed over its mouth, Enrique could see a hint of humanity in the creature's appearance. Its head was smallish and pear shaped. Its eyes were large, its skin was a faded pastel green and appeared to be textured in very fine and delicate scales. The creature was smaller than Enrique, about the size of four-year-old human.

As Enrique leaned forward to get a closer look, the alien opened its eyes. Although the movement startled Enrique, he immediately felt an overwhelming sense of compassion. Slowly, and with apparent difficulty, the alien raised a finger and pointed past Enrique. Instinctively, Enrique knew that someone had come down the stairs and was now behind him.

Enrique slowly turned, putting his hands in the air. Standing before him, holding a weapon that looked like a ray gun, was Agent Jones.

"This isn't what you think, Enrique," said the spook.

At just that moment, Captain Davis stepped out from behind the stairwell, a pistol in his hand.

"Roy, watch out!" shouted Enrique.

Agent Jones turned, but Captain Davis managed to fire his weapon. The bullet struck the spook squarely in the chest, knocking him to the floor.

"Thank goodness it's you, Roy," said Enrique. "It was Agent Jones, he was the real traitor."

Captain Davis stepped over the spook and collected the agent's futuristic weapon. Then …

"I'm afraid you're mistaken, Ric," said the boy's friend.

Captain Davis leveled the ray gun at Enrique and fired. A

brilliant stream of green light emerged from the ray gun and struck Enrique, sending him into spasms. Ropes of electricity danced over his body. He fell to the floor, smoke rising from his clothes.

Enrique struggled to remain conscious. His head spun and his body convulsed. Finally, Captain Davis took his finger off the trigger of the ray gun, approaching Ric with a distant, disturbed look in his eyes.

"Why … Roy … why?" croaked Enrique.

"Why, why? Because I'm better than this!" shouted Captain Davis.

"Why? Because the government, this great government of ours, used me! They used me and then they threw me in the gutter!" Captain Davis' face was now bright red; spittle flew from his mouth. His eyes were maniacal, the veins began to stick out of his neck.

"Why, you ask? It's because I was the best test pilot this stinking, no good, backstabbing government ever had. And what did they do? They grounded me! Me! How dare they! All because some desk jockey psychiatrist says I'm unfit for duty! They have me working with animals, Ric. Stupid, filthy animals! They want to train them for space. This great government of ours wants to send a monkey into space!! A monkey!! They were supposed to send me, Ric, me! I should be the first man in space."

For Enrique the room began to spin … *This can't be happening,* he thought. *Not Roy … the traitor can't be Roy.*

"But the Soviets and the commander, they recognized my brilliance … my skill … my genius!"

Captain Davis gazed upon the flying saucer with a longing look in his eyes.

"Well, now you know the secret, Ric," said Captain Davis in a soft, disturbing voice. "I see you've met the Nephilim. The alien crash-landed here in 1947. We've been trying to figure out the technology ever since, but we haven't even figured out how to open the craft. But when I deliver the Nephilim and its craft to the commander, I will be rewarded. Perhaps he'll even let me pilot it."

This must be a nightmare, thought Enrique. *It must be some terrible dream ...*

"I'm sorry it had to come to this, Enrique," said Captain Davis. "I really am."

Captain Davis raised the ray gun.

"Say hello to your mother."

At the mention of his mother, a spark of clarity lit Enrique. For a brief moment, the room stopped spinning and his stomach stopping roiling. In this moment of clarity, Enrique reached out and grabbed the pistol from the hand of the fallen soldier beside him. In one swift movement, he leveled the pistol and fired. Enrique saw the look of surprise on Captain Davis' face and then the room once again began to spin and all went black.

EPILOGUE

Enrique woke with the fading memory of the most pleasant dream he had had in some time. It was peaceful, very peaceful. He had been talking to his mother on the beach. They were laughing and playing … they were having a picnic … and his dad was there. His mother had entrusted him with something … it was the last thing she said … but as the clarity of wakefulness crept upon his mind, the dream began to fade. It was just before he opened his eyes that he recalled his mother's words …

"Take care of your father … he needs you."

Those were her words. The last words before his dream slipped away.

"Rise and shine, porcupine," said a woman's cheerful voice.

Enrique opened his eyes and his vision slowly came into focus. Leaning over him was the smiling face of Professor Appleby.

"We wondered if you were ever going to wake up," said the professor.

"Professor," said Enrique in a hoarse voice. "You're alive."

"Of course I am, Enrique. There's too much good music out there for me to up and leave just now, isn't there?"

Standing next to the professor were Carol and the general.

"Way to give us a scare, you doofus," said Carol with a playful smile on her face. Her braces made it so that doofus came out, "dufusssst."

"We found the professor in a hidden chamber behind

Captain Davis' laboratory," said the general. "I'm sorry for doubting you, son."

"What about my dad?"

"I'm right here, Mijo."

Enrique's father was sitting beside him, at the head of the bed. He had been holding Enrique's hand without the boy even realizing it.

"Your father has not left your side for the last three days," said the professor.

"I've been out for three days?" asked Enrique in a hoarse whisper.

"Ya sure have. And I've been waiting for ya to come to for the last day now." Tommy Reilly was propped up in the hospital bed next to Enrique. His face still had a purplish tint to it and he had a knot on his forehead, but he was all smiles.

"Tommy, you're okay!" said Enrique.

"Sure am. I guess that robot thing-a-ma-jig injected me with something it had cannibalized from the lab. The doctors are still trying to figure it out but they say I should be okay."

"The robot, what happened?" asked Enrique.

"Agent Jones had the foresight to install a hidden defense cannon near the saucer that would trigger if anyone — or thing — tried to access the craft," said the general.

"Agent Jones is … dead?" asked Enrique.

The general smiled and shook his head. "No, he's in intensive care, but he should be okay."

"But I saw him get shot in the chest. Point blank." said Enrique.

"He was wearing a vest," said Professor Appleby. "It's made from a new material called Kevlar. It's supposed to be bulletproof. I guess we haven't quite perfected it."

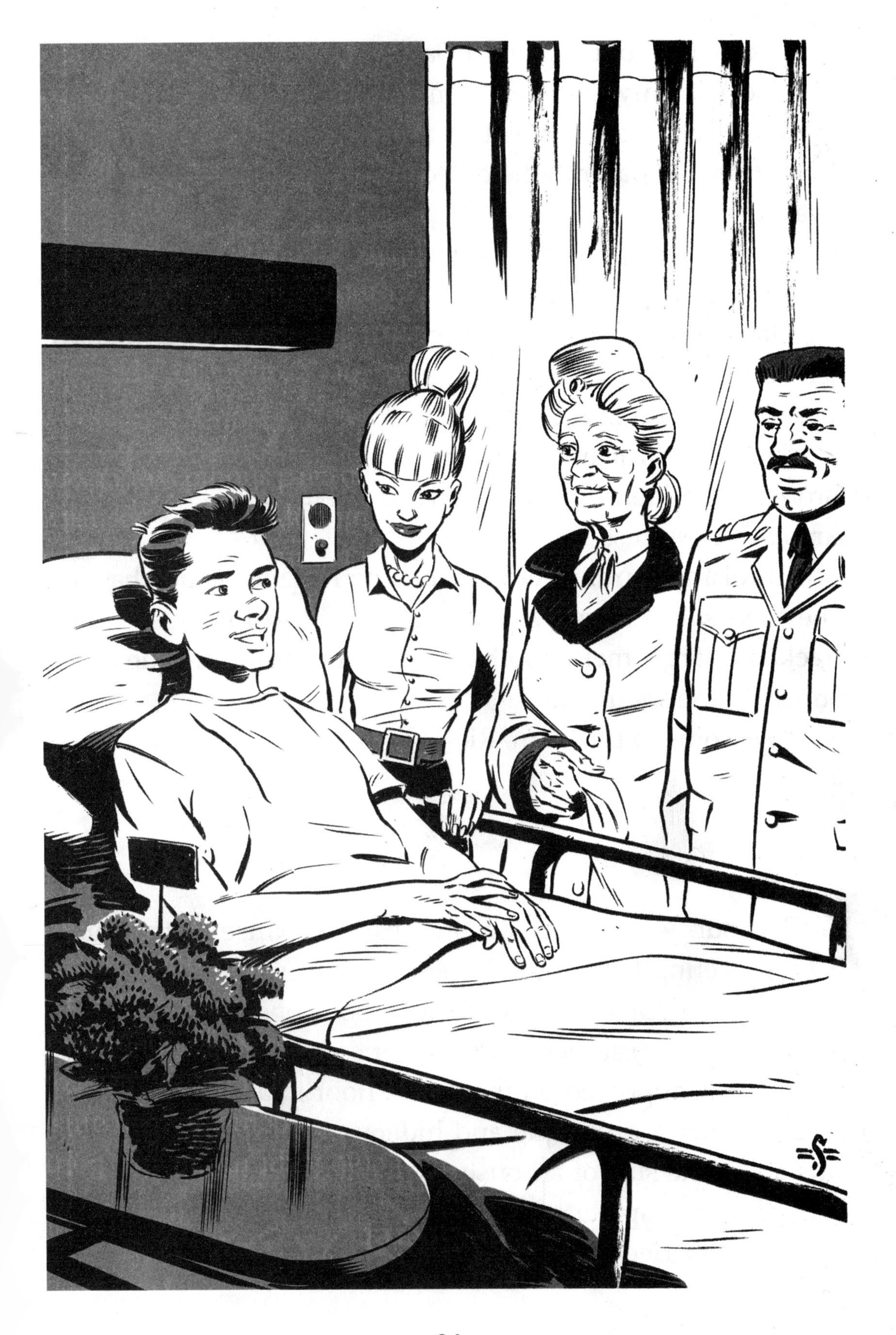

"But what about Captain Davis, did I kill him?"

"No, you only wounded him," said the general. "Currently he's in solitary confinement in the psychiatric ward awaiting evaluation. Apparently he defected to the Soviets a little over a year ago."

"What I saw ... in the vault ... Project Nephilim. Is it really an alien?" asked Enrique.

"It is," said the general. "And unfortunately, since you have been exposed to this highly classified and top secret program, there is only one thing for us to do."

The general leaned forward and handed Enrique a new Groom Lake identification card. It read, "Above Top Secret, Area 51, All Access."

Carol and Tommy held out their own identification badges. They, too, said, "Above Top Secret, Area 51, All Access." Carol cracked a large, metal-toothed smile and gave Enrique a soft punch on the shoulder.

"Welcome to the team, Ric."

The sun was setting on the desert expanse of the Yucca flats, reflecting brilliant splashes of light off the metal roofs and fences of Groom Lake Base. About one hundred yards beyond the nearest fence line, unnoticed by any human eyes, a small hole appeared on the desert floor in a puff of dust. A moment later a grotesque and hideous creature appeared. It was about the size of a house cat, its head humanlike and its body that of a spider. The creature expelled a vulgar cough and then scurried off across the lonely desert.

Past sage and cacti, past boulders and cliffs, the creature scurried well into the night. Past the town of Pahrump and past an unremarkable farmhouse and its tool shed. Behind the tool shed, it scurried down a hole.

Two hundred yards below the shed, Farmer Stefan Gelemne, A.K.A. the Commander, knelt within his foul garden. From a hole beside him the vulgar creature emerged, gasping for breath and clacking its hideous fangs.

"Ah, I see you have returned to me," said Farmer Stefan. "Excellent, I will be needing you soon."

Farmer Stefan picked the grotesque creature up by the scruff of its neck and placed it in an empty hole in the blood-red soil. The farmer then removed a scalpel from his lab coat and cut open his finger. As blood dripped into the soil, the multitude of grotesque creatures around him lifted their heads and wailed pleadingly.

The farmer held his bleeding finger over the open mouth of one of his creations.

"Yes, soon I will be needing all of you."

THE END

Watch for the return of Enrique Lopez in an all-new Cryptic Science *adventure! Now that Enrique, Tommy and Carol have Above Top Secret clearance, what further fantastic discoveries await them in the secret laboratories of Groom Lake Base? What sinister plans does the mysterious Farmer Gelemne have in store for the men and women of Area 51? And what fantastic secrets do the alien craft hold for our intrepid heroes? To find out, pick up the next edition of* Cryptic Science*!*

THE HISTORY

Stories of flying saucers and encounters with aliens have been common since at least the 1940s.

The most famous "real" story of a flying saucer occurred at Roswell, New Mexico, on July 2, 1947. That night, during a storm, witnesses claimed to have seen a saucerlike craft plummeting to the ground. The next day, a ranch worker came into the town of Roswell with strange pieces of metal – supposedly from the crash – which he gave to the town sheriff. Seeing that the debris had very unusual properties, the sheriff called the nearby Roswell Army Airfield to ask if anyone knew of any recent crashes. The base sent a major and a CIA officer to investigate.

Shortly afterward, the army swarmed in and removed truckloads of debris from the crash site. Local witnesses reported that the army took a damaged alien spacecraft and four alien bodies to hangar 84 at the Roswell Army Airfield.

Initially, the army released an official press statement declaring that they had captured a flying disk. The army later retracted this statement and declared the crash had only been a high-altitude weather-surveying balloon.

Later, the ranch hand, the sheriff and the investigating major changed their stories, all agreeing that they had not seen a UFO after all. Every speck of evidence was confiscated, and the military categorically denied the UFO reports. Rumors persist to this day.

Today, the public is still unsure of what really happened at Roswell. The military continues to deny the existence of any crashed flying saucer (they also deny the existence of Area 51). Roswell, New Mexico, however, has a booming tourism business and an alien museum dedicated to telling the story of the crash and the army cover-up.

THE PLACE – GROOM LAKE

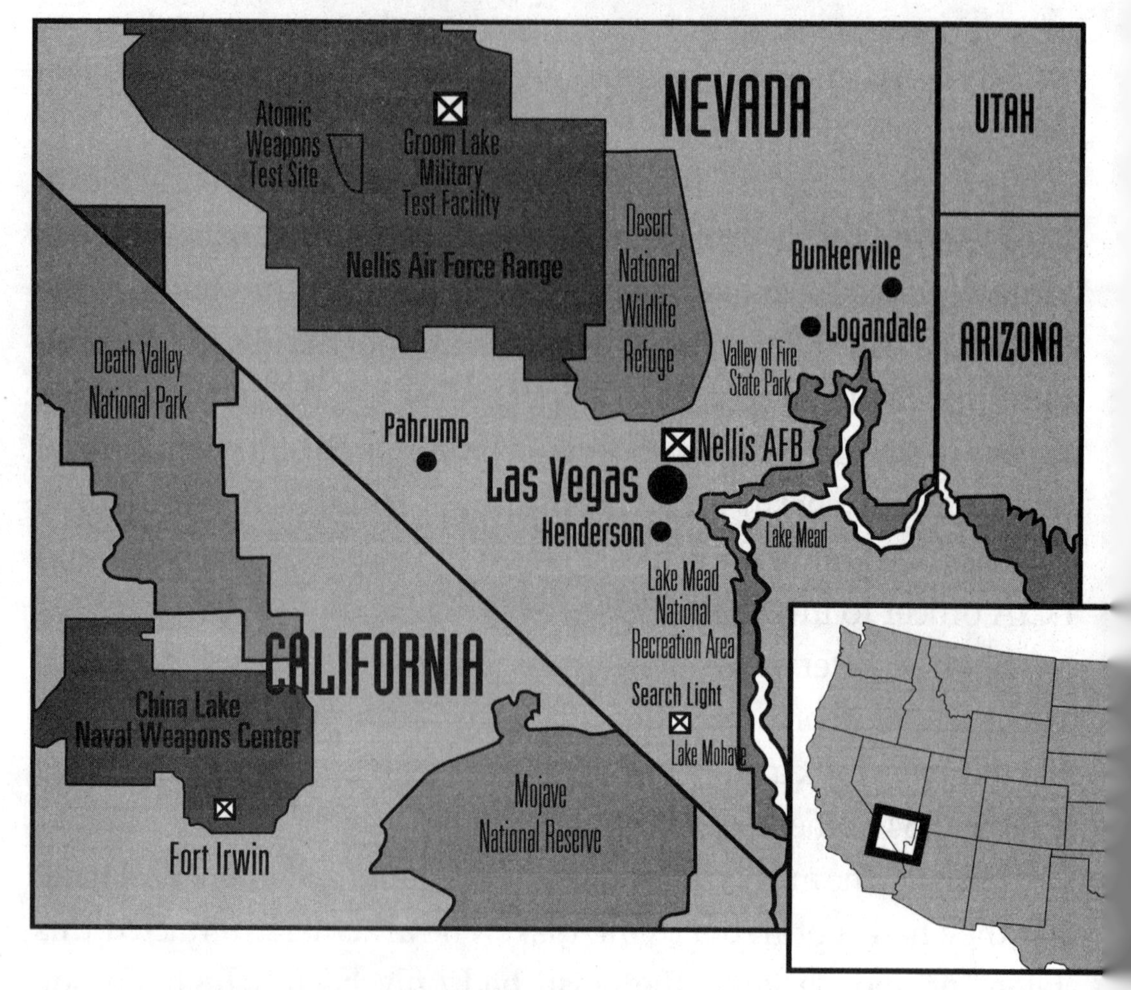

Area 51 does not exist. Not officially, at least.

According to modern satellite images and the majority of public opinion, Area 51 is located at the Groom Lake salt flats, in the Nellis Air Force Range in Nevada. This, supposedly, is where the government's most secret scientific work is conducted.

The Groom Lake salt flats were originally used as an artillery and bombing range during World War II. Then, in the mid-1950s, the area was converted into a secret test site for the U-2 spy plane. Many other state-of-the-art aircraft have been tested at Area 51, including the SR-71 Blackbird and the F-117 Nighthawk.

Many UFO enthusiasts are convinced that captured alien spaceships are hidden somewhere in an underground hangar in Area 51. The truth may never be known.

After all, the U.S. government has always denied the existence of Area 51.

PARTICLE RAY GUN

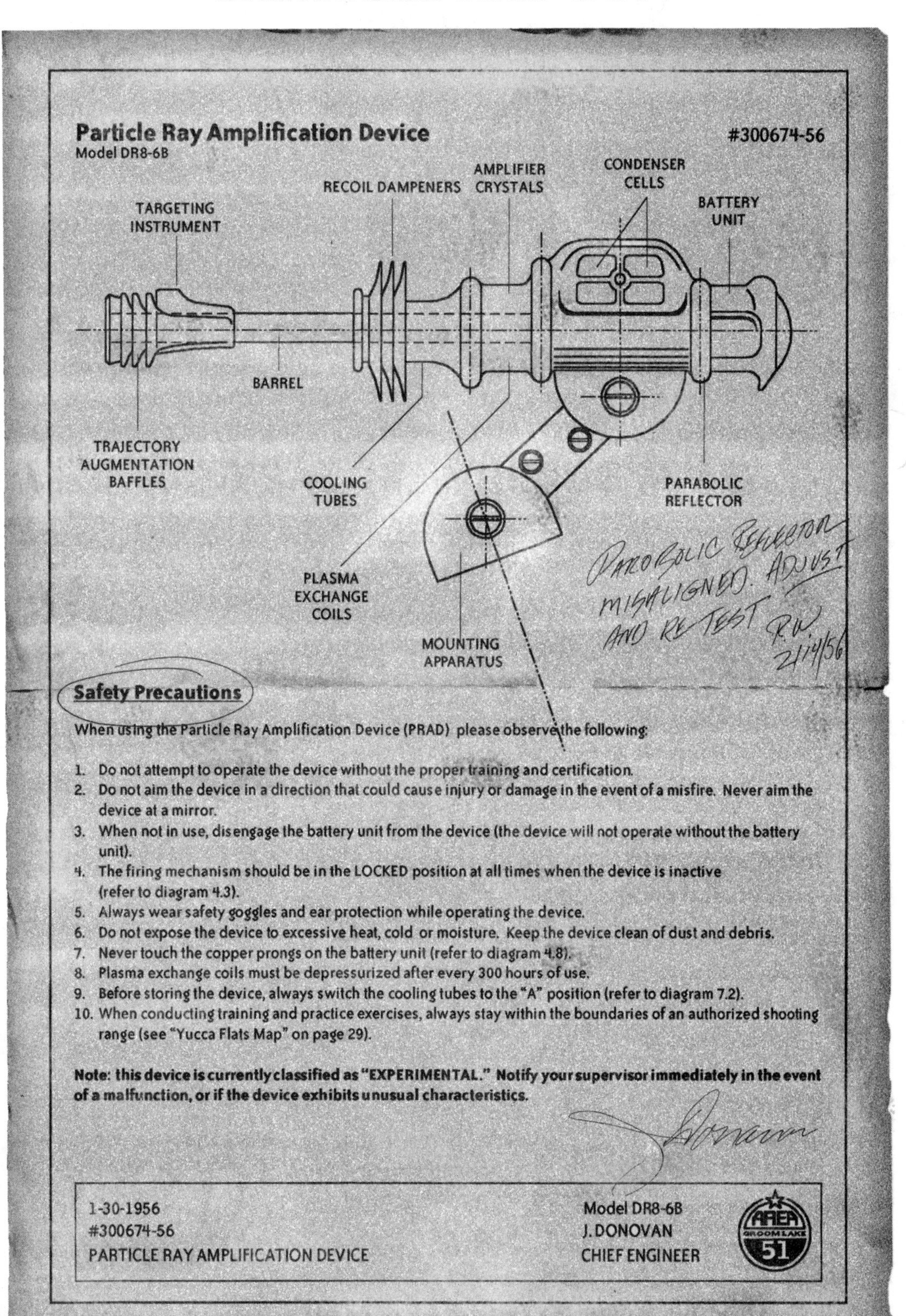

Particle Ray Amplification Device #300674-56
Model DR8-6B

Safety Precautions

When using the Particle Ray Amplification Device (PRAD) please observe the following:

1. Do not attempt to operate the device without the proper training and certification.
2. Do not aim the device in a direction that could cause injury or damage in the event of a misfire. Never aim the device at a mirror.
3. When not in use, disengage the battery unit from the device (the device will not operate without the battery unit).
4. The firing mechanism should be in the LOCKED position at all times when the device is inactive (refer to diagram 4.3).
5. Always wear safety goggles and ear protection while operating the device.
6. Do not expose the device to excessive heat, cold or moisture. Keep the device clean of dust and debris.
7. Never touch the copper prongs on the battery unit (refer to diagram 4.8).
8. Plasma exchange coils must be depressurized after every 300 hours of use.
9. Before storing the device, always switch the cooling tubes to the "A" position (refer to diagram 7.2).
10. When conducting training and practice exercises, always stay within the boundaries of an authorized shooting range (see "Yucca Flats Map" on page 29).

Note: this device is currently classified as "EXPERIMENTAL." Notify your supervisor immediately in the event of a malfunction, or if the device exhibits unusual characteristics.

Donovan

1-30-1956
#300674-56
PARTICLE RAY AMPLIFICATION DEVICE

Model DR8-6B
J. DONOVAN
CHIEF ENGINEER

AREA GROOM LAKE 51

GLOSSARY

Bill Haley and the Comets: *One of the first rock 'n' roll bands, most famous for their song "Rock Around the Clock." Their success paved the way for future rock musicians.*

Brylcreem: *A brand of styling mousse that gives hair a wet, shiny look.*

Dr. Josef Mengele: *A Nazi officer and physician at the infamous Auschwitz concentration camp during WWII. Dr. Mengele performed unspeakable medical and psychological experiments on prisoners. He often decided who would die and who would become a forced laborer. He escaped to South America after the war, avoiding punishment for his crimes.*

Gregori Yefimovich Rasputin: *Nicknamed "The Mad Monk," many people believed that Rasputin had unique supernatural abilities. They claimed that he had healing powers, psychic abilities and could see the future. He was a religious mystic who was oftentimes accused of misdeeds and wrongdoing. He was murdered in 1916.*

Handler: *In the world of spies and espionage, the handler is the individual who serves as a communications link between the spy and the agency or government that the spy works for.*

Humunculus: *A "false" human-being created in the test tube of an alchemist. Often a human, animal hybrid.*

James Dean: *A famous movie actor who starred in three of the most popular films of the 1950s (*East of Eden, Rebel Without a Cause *and* Giant*). Dean's life was cut short by a tragic car accident in 1955. He was only 24 years old.*

KGB: *Similar to the United States' CIA, the KGB operated as a secret police and intelligence agency from 1954 until the fall of the USSR in 1991. The KGB had a wide jurisdiction in Russia, dealing with spy and counter-spy operations, national security and many other issues.*

Sputnik 1: *The world's first space satellite, launched by the USSR on October 4, 1957. The launch of Sputnik marked the official launch of the space race between the USSR and the United States. Both countries knew that space exploration would lead to military and scientific advantages.*

Tsar Nicholas II: *The last emperor of Russia. He ruled from 1894 until the Russian revolution of 1917, when he was murdered by members of the Bolshevik political party (which later became the Communist party). In 1981, Nicholas and his immediate family were canonized as saints.*

Tunguska Event: *In 1908 a mysterious explosion occurred in the Krasnoyarsk Krai region of Russia which destroyed an estimated 80 million trees over 2100 square miles. Some attribute the massive explosion to a meteorite impact, while others believe the explosion was caused by a crashing UFO.*

ABOUT THE AUTHOR

A graduate of English from the University of Washington, S. Arthur Hart has been working as a Marketing Coordinator, game designer and copy writer for a small Seattle game company. He has authored numerous spec scripts for film and television and has written UFO book reviews for a quarterly new age periodical. Currently S. Arthur Hart lives in Seattle and spends his leisure time in the wilderness of the northwest conducting field research on that region's famously unidentified North American primate … Bigfoot.

ABOUT THE ILLUSTRATOR

Shane White *grew up in the Adirondack region of New York State and began drawing at an early age. He's worked in a wide variety of art fields for the last 15 years as an art director for a sculpting company, a 3D modeler for video games, a storyboard artist, concept designer and filmmaker.*

In 2005 he published his first graphic novel memoir, North Country, *and is currently writing and drawing the follow-up,* Things Undone. *He gets out of the studio once in a while to hike and paint the mountains surrounding his home in Seattle. For more information check out the following sites: www.shanewhite.com and www.studiowhite.com.*

Advancement in Business
Superiority in Sports
Flocks of Good Friends
Popularity with Others
Admiration and Respect
Are the best things in life passing you by?
Charles Atlas
Holder of Title "World's Most Perfectly Developed Man"
They Needn't Be — Because Now You Too Can Get A Big Brawny He-Man Body — the Kind That Helps to Bring Success in Sports, Business and Social Life
NO MATTER what good things may be passing you by—THEY NEEDN'T BE! Men with powerful HE-MAN bodies seem to have an irresistible appeal to others. They win the respect of men, the admiration of women. Their health, strength, and energy push them forward in life.
And right now, whether you're 16 or 64, skinny and weak or fat and flabby, you too can be a tower of strength and energy – with the kind of body you've always longed to have!
My Secret Formula
I used to be a 97-lb weakling. But now I don't have to take any back talk from anyone. I'm happily married, have lots of friends. I'm a success in business and social life. How did I do it? DYNAMIC TENSION – that's the ticket!
My secret formula, Dynamic Tension, wakes up sleeping muscles, shoots new life into them – makes them hard as tempered steel. It adds inches of firm rippling muscle to your chest, arms and legs – gives you broad, handsome shoulders that never fail to draw admiring glances.
You don't need any equipment or gadgets. Dynamic Tension requires only 2 simple things of you: (1) a little of your spare time; (2) willingness to build yourself into a better, stronger, healthier specimen of manhood. JUST 15 MINUTES A DAY–right in the privacy of your own room. That's all it takes – and soon you'll really begin to feel ALIVE, chock-full of zip and go!
TROPHY GIVEN AWAY Win this handsome Trophy, over 1½ ft. high!
JUST CHECK THE KIND OF BODY YOU WANT
More Powerful Arms and Grip
Broader Chest and Shoulders
More Weight --Solid in The Right Places
Slimmer Waist and Hips
Stronger Leg Muscles
Better Sleep, More Energy
CHARLES ATLAS, LTD. DEPT. AB2
PO BOX D. MAD. SQU. STA
NEW YORK, NEW YORK 10159
Name Age
(Please print or write plainly)
Address
City State Zip
Email
Dear Charles Atlas,
I am Sending in for my free information kit to find out more about your amazing "Dynamic--Tension" Course
1-888-Mr.Atlas ask for Operator AB2
1-888-672-8527 www.charlesatlas.com
© 2007 Charles Atlas Atlas Ltd.

THE ADVENTURE BOYS

James Tyler, Wild Boys Adventures™

James and his band of comrades fight for justice in the 1860s Wild West.

Sammy Wang, Detective Mysteries™

Sammy and his friends solve dangerous mysteries in 1930s San Francisco.

Ace Jackson, Blue Squadron™

Ace and his fellow pilots fly the skies of North Africa as they fight for victory during WWII.

Ric Lopez, Cryptic Science™

Ric and his band of young scientists battle the forces of evil at Area 51 in the 1950s.

Dusty Johnson, Spy Racer™

Dusty and his crew race for victory and battle Cold War spies in the 1960s.

Johnny Hawk, Treasure Raiders™

Johnny and his group of young archeologists encounter evil forces as they travel the world in the 1970s.